INTRODUCTION TO
ENVIRONMENTAL SCIENCE
LABORATORY MANUAL

KEVIN W. FLOYD | ELIZABETH J. WALSH

Kendall Hunt
publishing company

Cover image © Shutterstock.com

publishing company

www.kendallhunt.com
Send all inquiries to:
4050 Westmark Drive
Dubuque, IA 52004-1840

Table of Contents

Acknowledgments

The development of this lab manual was funded in part by a grant from the US Department of Education MSEIP program (P120A130103) "TIERA: Training in Environmental Research and Academic Success." Rip Langford and Joel Gilbert were involved in developing earlier versions of some of the labs for the Introduction to Environmental Science laboratory course. We thank the current and past Teaching Assistants for their input on developing and improving laboratories. Hundreds of student comments helped improve organization and clarity of labs, and provided the impetus to develop online supplementary materials. All figures and tables are created by the authors unless otherwise noted.

Introduction to Ecological Footprints and Sustainability

Learning Objectives

1. Investigate the impacts that personal and institutional choices have on the environment
2. Determine which decisions you make that have the largest impact on your ecological footprint
3. Consider the feasibility and importance of making both individual changes and institutional changes to reduce ecological footprints

Importance

Most actions that we take every day have some impact on the environment. Understanding the impacts may help people to choose actions that lessen the negative effect on the environment. Beyond personal choices, we are part of larger institutions such as the University of Texas at El Paso (UTEP), and the actions that administrators, facilities operators, purchasing agents, students, and others take also affect resource use. The human population is increasing, and as many parts of the world are experiencing economic growth, and personal consumption is increasing even more rapidly than population size. Accommodating increases in overall consumption of food, water, and energy while reducing cumulative environmental impacts is one of the greatest challenges we face today.

Introduction

The Industrial Revolution helped to improve food production, sanitation, and medicine. With a dramatic decrease in the death rates, the human population has sky-rocketed from around 1 billion in 1800 to over 7.3 billion in 2016. Along with the population increase, per capita consumption of resources has also risen. Use of natural resources, such as fossil fuels, water, and more recently, minerals that make up electronics, and the pollution generated by these uses, has placed incredible stress on all of the global ecosystems. We rely on these systems for climate regulation, water purification, pollination of food plants, and nutrient cycling, to name but a few. These are called ecosystem services, and although they are products of natural processes, humans depend on many of these services to maintain our quality of life. We thus face a great challenge in conserving healthy ecosystems if we are to maintain our current population growth rates and lifestyles.

One method to assess the impacts that humanity places on ecosystems is the ecological footprint. The ecological footprint, described in greater detail in the first lab in this unit, attempts to measure the amount of land and water required to provide the resources an individual consumes, and to absorb the wastes associated with that consumption. Associated with the ecological footprint are estimates of the renewable biological capacity of the various global ecosystems. Sustainable consumption would be equal to, or even less than, the biological capacity of the planet. However, recent estimates show that we are consuming resources at a rate 50% greater than the renewable biological capacity. This overconsumption is only expected to increase as the population grows past 10–11 billion by 2100, and as people in the developing world begin to consume more as their standard of living increases. We need to make changes in how and what we consume to bring our individual footprints in line with the Earth's biological capacity.

The first lab in this unit is an investigation of your individual ecological footprint. You will use an online calculator to first determine the footprint of your current lifestyle and then you will investigate what reductions are possible with different consumption choices. The second lab broadens the scope of the ecological footprint from the individual to the university. Universities are often major consumers of resources, and as such can have a large footprint. You and a partner will make a simple assessment of UTEP's ecological footprint, and then create options to help reduce UTEP's footprint. One of those options will be examined in more depth as you and your partner write a mock proposal to the UTEP Green Fund. The Green Fund is a program that helps to fund projects on campus that aim to reduce environmental impacts of various campus actions and increase the overall sustainability of UTEP. We will learn more about some of the aspects of the ecological footprint throughout the course, including carbon emissions from vehicles, water pollution, and sustainable food production.

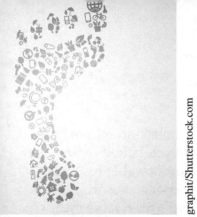

Lab 1

Individual Ecological Footprint

Learning Objectives

1. Explain how the concept of the ecological footprint is different than consumption of individual items
2. Classify activities into appropriate footprint categories
3. Calculate your ecological footprint
4. Investigate what types of changes have the greatest impact on your footprint

Importance

The environmental impact of our choices is often much greater than what we might expect. For example, when you buy an apple, the impact includes how that apple was grown (how much fertilizer and pesticides were used, was it grown organically, how much water was used in irrigation?), how it was transported to you (locally grown or shipped from across the United States or from across the world?), and what you do with it (do you eat it or throw it away after it begins to rot; do you compost the core?). The concept of ecological footprint attempts to include all aspects of the production, distribution, use, and disposal of all of the goods and services we use in our daily lives. The collective footprint of all humans on Earth is larger than the Earth's renewable resources by about 50%. Calculating your personal ecological footprint and investigating which decisions create the largest changes is an important step in increasing the sustainability of your lifestyle.

Guiding Questions

What everyday actions have the greatest impact on your ecological footprint?

What impact do large and occasional decisions (e.g., which house or car you might buy) have relative to everyday decisions?

How might increasing your knowledge about your personal footprint impact your future actions?

Vocabulary

1. *Global hectare*—Standardized unit that accounts for differences in the biological productivity of various ecosystems. There are about 16 renewable global hectares per person.
2. *Goods and services*—Goods are tangible items, while services are intangible and provided to a person (e.g., hospital care).

3. *Sustainability*—Generally, the ability to continue an action or behavior indefinitely. Environmental sustainability refers to the use of renewable resources below their rate of regeneration, the development of renewable substitutes for nonrenewable resources, and the creation of pollution at or below the rate that the environment can absorb and/or process them (Daly 1990). More simply, it is living within the Earth's renewable means.

Introduction

Ecological footprint estimates the total area of biologically productive land and water required to produce the resources that a person consumes, and to dispose of or recycle the waste that is produced (Withgott and Laposata 2015). The footprint is expressed in standardized units of area, either global hectares or acres. It can be broken into different categories, such as food or carbon. The footprint calculator that we will use in this lab has four categories: carbon, food, housing, and goods and services, briefly described as follows. Carbon footprint is based on the amount of carbon emissions caused by home energy use and transportation, which currently are largely fueled by fossil fuels. Food footprint includes the amount of cropland, pastures, and fisheries required to meet food consumption, along with the area required to absorb the carbon emissions associated with food production, processing, and transportation. Housing footprint includes the land that the home is on, along with the forests that produced the wood, cropland displaced by home water use (because that water is not available for irrigating crops), and the land required to absorb carbon emissions associated with construction and maintenance of the home. Finally, the goods and services footprint includes the land required to produce the raw materials of the goods (e.g., cell phone), to house commercial activities (e.g., store selling cell phone), and to absorb the carbon emissions associated with manufacturing, transport, and disposal of the goods. You can find more information about each category at http://myfootprint.org/en/about_the_quiz/faq/. Take a minute to think about an example of something you do every day that would contribute to your footprint for each of the four categories.

Carbon: _____

Food: _____

Housing: _____

Goods and services: _____

One key question is how much land can be used sustainably. The myfootprint.org calculator uses a recent estimate of about 16 global hectares (ha, about 40 acres) available per person on a renewable basis. Globally, humans use about 23 global ha per capita, or about 50% more than what is sustainable. We are overshooting the Earth's biological capacity, and would need 1.5 Earths to maintain current levels of consumption. Consumption varies globally, as discussed in the Sustainability chapter of your textbook, with people in the developed world typically having much larger footprints than those in the developing world. As the economies in the developing world expand and the global population continues to increase, humanity's ecological footprint will also grow. You might notice that the footprints reported in your textbook are much lower than the ones reported above. There are many ways of estimating the biological capacity of the planet, and the textbook uses a different method than the online footprint calculator we will use in this lab. Although the absolute numbers are different, both methods find that we are consuming the Earth's resources faster than they can be renewed.

It is easy to find lists of actions to reduce your ecological footprint online. Common suggestions are to drive less, turn off lights and electronics when not in use, and to recycle whatever can be recycled. These are relatively easy steps to take, but how much impact would taking these actions have on your footprint? Some actions that have a large impact on your footprint, such as not having children, not eating meat, or using only renewable energy, are more difficult for many people to do. Today's lab will help you explore these issues as you determine your current footprint and how much certain choices change your footprint.

Activities

You are going to calculate your ecological footprint three times using http://myfootprint.org, provided by the Center for Sustainable Development. Your TA will provide the username and password required to log into the site. There is a data sheet posted on Blackboard for you to record your answers for each question (see Table 1.1 for an example).

First you will answer all of the questions based on your *current situation and lifestyle*. At the beginning of the quiz, select "metric" for the units. You can see your current footprint and the country average footprint in a small box at the top of the webpage. As you answer each question, you will see the current footprint change. Record your answer(s) for each question on the data sheet in the column labeled "Original footprint" (see Table 1.1). At the end of the quiz, you will see how many Earths would be required if everyone on the planet lived your lifestyle, and how your footprint in each of the four categories compares to the national average (Fig. 1.1). Record those results on your lab assignment.

Second, you will go back to the beginning of the quiz (use the back button at the bottom of the webpage whenever possible). This time you will answer the questions based on what actions you could *realistically accomplish*. For example, it might not be realistic for you to buy a hybrid vehicle, but it might be realistic for you to carpool to UTEP once a week. Record the changed answers on the data sheet (Table 1.1). If there is no change, write "same." Once you complete the calculator, record the new footprint on your lab assignment, and calculate the difference between your initial footprint and the new footprint.

Third, you will go through the quiz a final time, but this time making sufficient changes to *reduce your footprint by 50%* in each category. First you will calculate how many hectares your reduced footprint will be, and record the results on the lab assignment. Then go back to the beginning of the quiz and start making changes that you think will reduce your footprint

Table 1.1: Example of the data sheet for recording your answers on the footprint calculator. The original footprint is for your current footprint. The next two columns are for recording changes to your original answer. If you cannot make a change for that question, enter "same."

Question	Original footprint	Changes you think you can realistically make	Changes required to reduce original footprint by 50%
Carbon footprint			
5) Climate zone	Warm to hot lowland desert	same	same
6) Size of home	150–200 sq meters	same	same
7) Energy sources	Electricity, natural gas	same	same
8) Percentage of electricity from renewables	5	same	same
9a) Kilometers per year by car	16,000	13,000	5000
9b) Type of car	mid-sized car	same	Hybrid

If everyone on the planet lived my lifestyles, we would need:

= 5.13 Earths

Petitions by change.org. Start a petition

Reduce your footprint

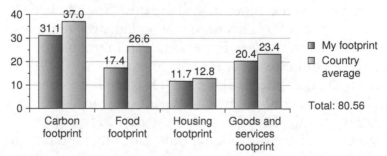

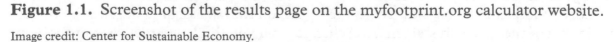

Figure 1.1. Screenshot of the results page on the myfootprint.org calculator website.

Image credit: Center for Sustainable Economy.

in that section. Use the box at the top of the webpage to see how large of an impact each change has on your footprint. Keep making changes until your footprint is only half of what it was originally. Record your answers in the last column of the data sheet (Table 1.1).

Finally, you will choose one change in each of the four categories to try for a week. They can be substantial, such as eating only organic food, or they can be smaller, like running the dishwasher only when it is full. The important thing is that they are changes that you think you can actually make during the week. For each change, you will need to also explain how it actually reduces your footprint and how feasible it is both financially and logistically. You will answer these questions on the lab assignment. Do not get discouraged if you find it more difficult than you expected to make the changes. Our behaviors are often difficult to change. Just keep trying, and ask your friends and family to join in for encouragement and support.

Assessments

Complete both the data sheet including all three times that you go through the calculator and the lab assignment. Both are available on Blackboard.

References

Center for Sustainable Economy, n.d. "Ecological Footprint: What it Measures." http://myfootprint.org/en/about_the_quiz/what_it_measures/. Accessed October 29, 2016.

Daly, H. 1990. "Toward Some Operational Principles of Sustainable Development." *Ecological Economics* 2 (1): 1–6.

Withgott, J. and M. Laposata. 2015. *Essential Environment: The Science behind the Stories*. 5th ed. San Francisco: Pearson Education.

Resources

The FAQ for the footprint calculator has more details about each of the categories: http://myfootprint.org/en/about_the_quiz/faq/.

The Global Footprint Network has a lot of detailed information for many countries: www.footprintnetwork.org/en/index.php/GFN/.

National Geographic produced a video called "Human Footprint." Clips are available online, and they have a really nice interactive page with the footprint of several everyday items: http://education.nationalgeographic.org/interactive/human-footprint-interactive/

© K. Floyd

Campus Ecological Footprint

Learning Objectives

1. Assess environmental impacts of the UTEP campus
2. Classify UTEP campus activities into the four footprint categories
3. Identify areas for environmental improvement

Importance

College campuses use many resources, resulting in large ecological footprints. However, higher education also has generally moved towards increased sustainability through individual campus actions and broader commitments such as the American College and University President's Climate Commitment. One sustainability initiative at UTEP is the Green Fund, which dispenses monies to fund projects around campus that will lessen our collective environmental impact. Assessing UTEP's ecological footprint will help students to gain an understanding of how the UTEP campus is impacting the environment and to generate student ideas for new Green Fund projects.

Guiding Questions

Should UTEP take more actions to increase campus sustainability?

What campus activities have the greatest environmental impacts?

What reasonably scaled projects could have a measurable impact on UTEP's sustainability?

Vocabulary

1. *Sustainability*—Generally, the ability to continue an action or behavior indefinitely. Environmental sustainability refers to living within the Earth's renewable means.

Introduction

College and university campuses have classrooms, offices, research labs, food services, student housing, and sports arenas and stadiums, among other facilities. Running campuses requires large amounts of energy, water, food, and supplies such as computers, paper and ink, and produces a variety of wastes. Universities are often one of the largest businesses in their communities. It is estimated that UTEP represents about 5% of El Paso County's total economy, a larger share than construction (UTEP 2014). UTEP spent $202 million on payroll and operations in 2013, and has built new facilities costing $300 million over the past five years (UTEP 2014). Decisions about how UTEP builds and operates have large environmental consequences. Designing new buildings and retrofitting existing buildings to be efficient in both energy and water use will decrease both UTEP's ecological footprint and continuing operation costs.

In addition to the economic importance of campuses, they also can serve as laboratories for testing sustainability initiatives (Fihlo et al. 2015). Universities tend to have more flexibility and freedom to try new ideas than do businesses that need to report profits to shareholders. Many universities have successfully involved students in sustainability projects, providing real-world experiences in skills that future employers find valuable (Fihlo et al. 2015). At UTEP, if students, faculty, or staff have ideas for improving sustainability, they can submit a proposal to the Green Fund Committee (information at http://sa.utep.edu/greenfund/). The Green Fund was created in 2010 following a student-led initiative, and uses charges of each student $3 per semester, raising up to $40,000 per year to support sustainability initiatives (http://sa.utep.edu/greenfund/). Funded projects have included a food garden at Miner Heights student housing, the Miner Recycling System currently in place at the Union, Business, and Health Science buildings, and solar panels for Miner Heights (http://sa.utep.edu/greenfund/projects/). After this lab, students will work in pairs to develop a modified Green Fund proposal to improve sustainability at UTEP. Details about the specifics of the proposal are given after these lab instructions.

To help generate ideas for the proposal, we will conduct a simple environmental audit of UTEP. This audit will focus on the same four categories as were included in calculating your individual ecological footprint: carbon, housing, food, and goods and services. Each group will visit two buildings and look for items related to energy use (carbon footprint), food (food footprint), water and cleaning products (housing footprint), and materials such as paper, computers, and furniture (goods and services footprint). Each group will also look at the landscaping, focusing primarily on water and waste management, and food options at either the Union Food Court or the El Paso Natural Gas Center. Finally, each group will look at transportation options, which also directly impact UTEP's carbon footprint. There is a list of questions for each campus area in the lab assignment (available on Blackboard). It might not be easy to find an answer to every question, which leads to another series of thought questions: How transparent should UTEP be about how it operates? Should information about cleaning products, waste management, and fuel use be readily available to UTEP students, or the general public? If so, how much would it cost UTEP to track and publish such information?

It is important to consider the type of data that we will collect. Scientists usually collect three types of data: discrete, continuous, and categorical. Discrete and continuous are both numerical data. Discrete data are items that are often counted and can only be fixed values, such as the number of trees on campus or students in Environmental Science. Continuous data are measurements that can be any value, such as the height of the trees or the amount of carbon dioxide emitted from a car. Categorical data represent distinct characteristics, such as sex (male or female), habitat type (forest, desert, or marine), or type of rock (igneous or sedimentary). Although we could collect discrete data (how many computers are on in the library?) or continuous data (how much fuel do UTEP fleet vehicles use daily?), our main goal is to get an estimate of each category. So you will put the numerical data into categories such as "very little," "some," or "lots". Typically scientists record numerical data because it is more accurate and does not depend on a person's individual opinions. For example, you might think that UTEP has a lot of recycling bins on campus, but another person might think that there are not enough. Counting the number of recycling bins provides data that everyone can agree on. As you work on your Green Fund proposal, think about collecting numerical data whenever possible.

Activities

In pairs, complete the campus environmental audit. Your TA will assign each pair two buildings: one will be relatively new and the other will be older, and you will look for differences between the two. You will also be assigned either the Union Food Court or the Natural Gas Center to assess the food options. As you walk around campus, assess both the landscaping and the transportation options. Every group will give a short presentation on their findings when they return to the lab. For homework, you will research the specific environmental impacts for each of three issues that you observed. One of these will be the basis for your Green Fund proposal.

Assessments

Complete the lab assignment available on Blackboard.

References

Fihlo, W. L., Shiel, C., Paco, A. do., and Brandli, L. 2015. "1—Putting Sustainable Development In Practice: Campus Greening as a Tool for Institutional Sustainability Efforts." In *Sustainability in Higher Education*, edited by J. Paulo Davim, 1–19. Amsterdam: Chandos Publishing.

University of Texas at El Paso (UTEP). 2014. *Economic Impact.* http://impact.utep.edu/impact.html. Accessed October 29, 2016.

Resources

UTEP's sustainability webpage has some information about campus actions, particularly about the Miner Recycling Program: http://gogreen.utep.edu/

The Association for the Advancement of Sustainability in Higher Education (AASHE) has a lot of resources about sustainability on campuses. In particular, look at the Resources for Students (http://www.aashe.org/resources/general-resources-campus-sustainability/student-resources) and the Campus Operations Resources (http://www.aashe.org/resources/campus-operations-resources/) for ideas of how to improve UTEP's sustainability and for ideas for your Green Fund proposal.

Instructions for the Green Fund proposal

You and your partner will write a mock proposal for a sustainability project to be considered for funding by the UTEP Green Fund. You should develop one of the ideas generated during the Campus Ecological Footprint lab. A mock proposal is one that contains all of the elements required by a funding agency but is not actually submitted for formal evaluation. Information about the Green Fund can be found at http://sa.utep.edu/greenfund/.

The goals for this assignment are for you to (1) learn about sustainability, (2) think critically about how to improve sustainability of campus activities, (3) consider the potential conflicts between environmental and economic/societal concerns, (4) develop feasible solutions to problems, and (5) learn to prepare a succinct and compelling funding appeal. Notice that this is not a strictly scientific endeavor: part of environmental science is the identification of problems and determining potential solutions, but actual implementation of changes often depends on economic and societal costs and benefits. You and your partner will likely have diverse backgrounds. The interdisciplinary nature of improving sustainability of campus activities will benefit from the unique knowledge each of you will contribute.

Check on Blackboard for the specific due dates and the grading rubric for the proposal. You will turn in a first draft of your proposal in a few weeks. This needs to be a complete draft written to the best of your abilities. Before turning it in, the draft must be reviewed at the UTEP writing center (http://uwc.utep.edu/). They will help with grammar, spelling, and overall writing quality. You must turn in proof that you received help with the final draft. You will receive another group's proposal to review following the same grading rubric. The peer review itself will be graded, and will be due the week after you receive the proposal to review. After receiving comments and making corrections and improvements, your final proposal will be due around the tenth week of the semester. It is highly advisable to get it reviewed at the writing center one more time before turning it in. Review the rubric and the following instructions as you work on your proposal.

Remember that this is a proposal for funding. You are asking the Green Fund committee to fund your project instead of other proposals. You need to make a *strong case* for the importance of your project, and you also need to convince the committee that what you are proposing is both feasible and beneficial. Remember you are not required to actually submit your proposal, but if you are excited about your idea and would like to submit it, we can help you begin the actual submission process.

Please follow these guidelines to ensure that you receive the maximum grade possible. Each section should be one or two paragraphs, so the overall proposal should be two or three pages. Also check the grading rubric posted online before preparing your proposal so you fully understand the expectations of this assignment.

Campus Sustainability

Describe the problem

Based on your campus environmental audit from the Campus Ecological Footprint lab, what is the problem you will address? What are your observations about the problem based on the audit and other information you have gathered on the topic?

Describe the negative impacts

What are the specific impacts of the problem to the environment, the campus, and the economy/society? You will need to do some background research here to be able to convince the committee that this is a serious problem that needs addressing. You will need to include references (described below).

Proposed solution to the problem

Describe the proposed actions: Give specific actions that will be taken to address the problem you described. The more specific you can be, the easier it is for the Green Fund committee to understand your proposal.

Describe how the impacts of the action will be measured: It is important to assess how well the actions work. This includes logistics (e.g., are recycle bins being used correctly?) and overall impacts (e.g., how many pounds of plastic are being collected?).

Describe the publicity/outreach activities: It is important that people both at UTEP and in the community know about the actions. This could be instructional (e.g., signs stating the types of materials that can be composted) or educational (e.g., reporting the amount of energy savings by using LED instead of incandescent bulbs), but it is crucial to let people know why this activity is important to help sell your idea.

Benefits to the Environment and UTEP

Specific benefits to the environment

Connect the actions to the environmental problem you identified in the beginning. How is the condition of the environment benefited by your proposed actions? Can you quantify these benefits?

Specific benefits to UTEP

Connect the actions to improvements to UTEP. Will it improve the environmental surroundings or health of the UTEP community? Will it save UTEP money?

Specify whom the project will serve and/or benefit

Who are the people served and/or benefited? Is it everyone, or just a subset of the community?

Feasibility

Rough estimate of the costs

This can potentially be difficult if you get into the details of every aspect of your proposed actions. You can keep it general, such as how expensive solar panels are, how much LED bulbs cost, and so on. You need to identify both one-time and on-going costs.

Estimate of logistics

Will the proposed actions be easy or hard to implement? Do you anticipate the administration being resistant or eager to implement your actions? Will the students, staff, and faculty be resistant or eager?

Anticipation of problems

What do you think could go wrong with implementing your proposed project, and what actions can be taken to overcome the problems? This part demonstrates that you have thought deeply about the actions and how implementation is likely to occur.

Conclusion

Summarize the problem, the solutions, and the benefits

The overall proposal should follow a hero narrative: Here is a big problem. We have the solution. Here is how the solution will benefit the environment, UTEP, and society. The conclusion is the final take home summary of everything included in the proposal.

Literature Cited

You must cite at least **five** references in your report. These can be your textbook or websites. Be sure to use only reputable websites (those associated with journals, university professors, etc.).

You will need to do some research to write a successful proposal. You need to include **both** *in-text citations* and a *literature cited section* at the end of the report. Whenever you use a reference for factual information in your report, cite it. Avoid using direct quotes, even with proper citations. Instead, paraphrase the important information and cite the reference. The format to use is parenthetical: (Author date).

For the literature cited section, you need to use consistent and formal formatting. It can be APA or MLA if you have learned them in other classes or you can use a different format with which you are most comfortable. The focus for this class is learning the basic concepts of how to use and cite references correctly, not on specific formats, so you can format however you prefer as along as it is consistent for each reference. For information on citing websites, go to http://www.studygs.net/citation.htm for instructions. *You cannot simply give website URLs*—you need to include the name of the site, the URL, and the date visited at a minimum. There are many online resources on how to format literature cited sections.

General Instructions

The proposal must be typed and double-spaced. Proofread it at least once before taking it to the writing center. Use the spell- and grammar-checker. Include the section headers to help keep your thoughts organized. Proofread it again, focusing on the logic and flow of the proposal.

This proposal is a major part of your grade, which means that we take it very seriously and expect you to do the same. Your TA is always willing to help, but you need to ask. The grading rubric that we will use to evaluate your final report is posted on Blackboard. Use it to help guide you in constructing your report. Hint: Make sure that you include all of the components that will be evaluated to ensure that you receive full credit.

Have fun! This is a real-world activity, and we encourage you to consider submitting a proposal to UTEP's Green Fund committee.

Introduction to Population Growth

Learning Objectives

1. Understand processes that cause populations to change in size
2. Connect resource availability to population growth
3. Discuss costs and benefits of human population growth
4. Relate the average ecological footprint of various countries to the potential human carrying capacity
5. Use the algae biofuel experiment to explain a practical application of relationship between resource availability and population growth
6. Apply principles of the scientific method to design an experiment and interpret the results

Importance

Understanding the factors that cause changes in population sizes is fundamental to most environmental concerns. The number of births, deaths, immigrants, and emigrants determine population size. Researchers study many types of populations to understand what influences rates of births and deaths, as well as those for immigration and emigration. Perhaps the most important population to consider is that of humans. As the number of people increases, the amounts of resources such as food, land, and water required to support them also increases. Predictions that humans would run out of these resources have not yet come true, largely due to unanticipated technological advancements in water treatment and crop production, among others. Projecting human population growth accurately is critical for planning how to sustainably manage natural resources and has major implications for population sizes of other organisms.

Introduction

Populations increase when new individuals are born or immigrate into the region, and decrease when individuals die or emigrate out of the region. Birth rates and death rates for most organisms are directly related to resource availability: as resources become more limited, birth rates decrease and death rates increase. Migration rates also depend on resources: as resources become more limited, more individuals leave the population than enter. When the numbers of individuals entering and leaving the population are equal, the population is in equilibrium. The carrying capacity of an environment is the maximum number of individuals of a given species that can be sustained on the available amount of resources. It changes when the amount of resources change, increasing during times of resource abundance and decreasing during times of scarcity.

Questions about the carrying capacity of humans have been asked for several hundred years. One camp predicted that humans were going to reach or even overshoot our carrying capacity in the near future. This has yet to happen, largely due to technological advancements such as the Green Revolution in agriculture that effectively increased the availability of resources, as well as the development of a global food distribution network. Some researchers think that human ingenuity will continue to remove resource-related limits to our population size, and that the human population essentially has no carrying capacity.

The question of what is the human carrying capacity is difficult to answer, in large part because it depends on what standard of living we want for everyone and how much of the natural world we will protect from exploitation. The human carrying capacity would be much larger if everyone had an ecological footprint similar to that of an average Kenyan. If everyone wants to live the lifestyle of an average American, the increased consumption of resources means that the carrying capacity will be much lower. Our environmental impact is an interaction between individual resource use and the total number of people. Although most environmental scientists would likely agree that environmental degradation would be minimized with a smaller human population, there are social and economic reasons for governments to encourage continued population growth. Human population growth is a fascinating and complex topic.

In the first lab in this unit, you will learn more about the environmental, economic, and social issues related to population growth in three contrasting countries: India, Japan, and Kenya. You will also investigate the relationship between population growth rates and the time that it takes for a population to double in size. In the second lab, you will investigate the impact of resources on population growth using algae as a test population. Students will collaborate to test how the concentration of two key nutrients, nitrate and phosphate, affect population growth of the algae. Understanding how to grow large populations of algae is important for the production of biodiesel, an alternative transportation fuel. Currently, we rely on oil to power transportation, but the combustion of oil releases carbon dioxide, one of the most important greenhouse gases and the main human-caused gas contributing to climate change. The algae experiment will also give you an opportunity to practice all the steps of the scientific method, from formulating a hypothesis to data analysis to drawing conclusions. At the end of this unit, you will understand how resources impact population growth, the challenges and opportunities created by human population growth, and how to conduct a scientific experiment.

Arthimedes/Shutterstock.com

Lab 3

Human Population Growth

Learning Objectives

1. Explain the differences between exponential and logistic population growth
2. Understand how current birth rates affect future population sizes
3. Compare challenges and opportunities caused by high population growth rates to those caused by low population growth rates
4. Distinguish between environmental and economic costs and benefits

Importance

Population growth is fundamental to many concerns in environmental science. Conserving endangered species requires understanding what factors are leading the population to decline. Understanding what causes particular species to have their populations increase is critical in slowing the spread of invasive species. Understanding how human populations grow is necessary to determine how much growth is expected over the next century, and what measures can help to slow growth. Although people in different countries consume resources at different levels, and reducing individual consumption can help to reduce the overall ecological footprint, many researchers agree that an increasing population puts additional strains on the planet's resources. In addition, social, political, and economic pressures come into play as human populations grow. Investigating the challenges and opportunities that various levels of population growth present to countries will help you to appreciate the complexities surrounding human population growth.

Guiding Questions

Under what conditions would populations grow exponentially?

What are the key resources that could potentially limit the human population?

Does the human population have a carrying capacity?

How should individuals and governments balance the costs and benefits of population growth?

Vocabulary

1. *Population*—A group of interacting members of the same species in a given area.
2. *Population density*—The number of individuals per unit area (for terrestrial organisms) or unit volume (for aquatic organisms).
3. *Population growth rate*—The change in population size during a period of time, usually expressed as a percentage of the population at the start of the period.
4. *Migration*—Movement of individuals into a population (immigration) or out of a population (emigration).
5. *Density-independent population growth*—The type of population growth that results from a constant population growth rate across all population densities. Also called exponential or J-curve growth.
6. *Density-dependent population growth*—The type of population growth that results from a population growth rate that changes as the population density changes. Also called logistic or S-curve growth.
7. *Carrying capacity*—The population density that an environment will sustain indefinitely. The carrying capacity changes as the environment changes, such as during droughts or after fires or as resources are depleted.
8. *Total fertility rate (TFR)*—The average number of offspring per female.

Introduction

Many questions in environmental science require understanding population dynamics. How will urbanization or pollution impact local wildlife populations? How can we increase the populations of fishes caught for food? What is causing the populations of some species to decrease while others increase? What factors influence the growth of the global human population? What might lead to the leveling off and eventual decrease in the human population?

Basic Population Growth

Populations increase when new individuals are born or immigrate into the population, assuming no changes in death or emigration rates. Likewise, populations decrease when individuals die or emigrate from the population when birth and immigration rates are stable. The balance between these four factors determines whether populations increase, decrease, or remain constant. The number of births, deaths, immigrants, and emigrants can potentially be monitored in a small population, but for larger populations, a subset of individuals is studied and per capita (per individual) rates are calculated. In studies of human populations, you might see *crude* birth, death, or migration rates, which are the numbers of births, deaths, or migrants per 1000 individuals. The overall population growth rate is the change in the population size from the beginning to the end of a particular time period, and is equal to the increases from births and immigration and the decreases from deaths and emigration. This is expressed as:

$$N_{t+1} = N_t + B + I - D - E,$$

where N_t is the population size at the beginning of the time period and N_{t+1} is the population size at the end of the time period. The length of the time period varies based on the organism and research question. When studying bacteria, the time period might be an hour, while for humans it might be a year or a decade. Although migration is important for many research questions, we will ignore it in these labs for simplicity. Thus, the population growth is simply the balance between births and deaths. The per capita population growth rate, r, is equal to the per capita birth rate, b, minus the per capita death rate, d ($r = b - d$).

Types of population growth

When the per capita growth rate remains constant across all population sizes, the population grows **exponentially** (Fig. 3.1A). When the population is small, the constant growth rate produces a small increase. But each of those new individuals then contributes to the population growth, so the population increases by a larger number of individuals as time progresses. Assume a per capita growth rate of 5% per year. Calculate the number of individuals added in year with the different starting population sizes by multiplying the starting population size by the growth rate.

Starting population size	Number of individuals added	Ending population size
10		
100		
1,000		
10,000		

How did the number of individuals added change as the starting population size changed?

_____

_____

How does this pattern fit the exponential growth curve shown in Figure 3.1A?

_____

_____

One related concept to exponential population growth is *doubling time*, which is simply the amount of time that it takes for the population to double when the growth rate remains constant. You will be investigating how population growth rates impact doubling time in the activities for this lab.

Exponential population growth is also called **density-independent** population growth because the growth rate is the same at all population sizes, small and large. Population density is simply the population size per unit area (or unit volume for aquatic organisms). Individual organisms require resources to survive, grow, and reproduce. For a population to grow exponentially, individuals must be able to obtain all of the required resources at the same rate, whether there are 10 individuals or 10,000. This rarely happens for extended periods in nature because some resource is likely to be in short supply, and individuals will begin to compete for such resources as the total number of individuals increases. For example, black-tailed jackrabbits near El Paso might not compete for food when there are only 10 individuals per hectare, but would definitely compete for food if there were 1000 individuals per hectare. With increased competition for resources, individuals often produce fewer offspring and the probability of dying increases. Decreased reproduction and increased mortality both reduce population growth. Because of this, the population growth rate slows as the population density increases, eventually reaching a point when the birth rate and death rate are equal ($b = d$) and the population stops growing (Fig. 3.1B). This type of growth is called **density-dependent** and follows the **logistic** growth model. The population size when the birth rate and death rate are equal is called the **carrying capacity** (abbreviated K). The carrying capacity depends on the environment, and changes when the environment changes. For example, if there is above-average precipitation several consecutive years, the increased plant growth will be able to support larger populations of jackrabbits. However, if there are several consecutive years of drought, the decreased plant production will support a smaller population of jackrabbits. The carrying capacity is the sustainable population size. Populations can overshoot their carrying capacity, but will then experience a death rate higher than the birth rate ($d > b$) and the population will decrease to the carrying capacity.

Density-independent growth assumes that resources are unlimited, and the population will be able to grow at its biological capacity. Density-dependent growth recognizes that some resources are limited, and that populations will grow until those resources limit further growth. Such resources are called **limiting factors**, and for most species, there will be just a few resources that limit the population size, regardless of the availability of other resources. For plants, nutrients such as nitrogen and phosphorus are often limiting, while the sunlight and carbon dioxide required for photosynthesis typically are plentiful. For some animal species, the amount of food might be limiting, while for others the number of suitable

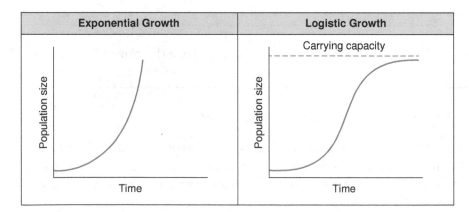

Figure 3.1. (A) Exponential population growth, also called density-independent or J-curve. The population growth rate remains constant across all population sizes. (B) Logistic population growth, also called density-dependent or S-curve. The population growth rate decreases as the population reached its carrying capacity.

SOURCE DATA from https://cnx.org/contents/eeuvGg4a@4/Environmental-Limits-to-Popula#fig-ch45_03_01.

nesting locations might limit the population size. Researchers often try to determine what limits the growth of population for a particular species. If it is a species that people want to harvest, then understanding what resource limits growth would allow someone to increase that resource. For example, a farmer can increase the amount of nutrients if that is what limits the growth of a crop. You will investigate how nutrients impact the growth of algae in the next lab in this unit.

Human population growth

The global human population has grown dramatically over the past 100 years. The global population reached 2 billion in 1927, 3 billion 32 years later, 4 billion 15 years after that, with an additional billion people about every 12–13 years (Withgott and Laposata 2015). During this time, the population grew faster than exponential, with the per capita growth rate increasing until the 1960s. The growth rate has decreased since then, but the large population ensures a large number of new individuals each year. The current global population is over 7.3 billion people, with about 80 million added each year. You can see the current population size at http://www.census.gov/popclock/.

Human population growth was largely fueled by technological advancements that decreased the death rates, such as antibiotics, sanitary disposal of waste, and increased food production. Birth rates typically do not decrease as rapidly as death rates, so overall population growth increases. In most parts of the developed world, birth rates eventually decrease and population growth rate slows. However, many developing countries are currently in a growth period, and much of the population growth projected for the 21st century is expected in those countries, particularly in sub-Saharan Africa. The size of the population in 2100 will largely depend on what happens to the birth rates globally. To maintain a constant population size, the total fertility rate (TFR) is 2.1 children. The two children replace the parents, and the extra 0.1 represents the number that might die before reaching reproductive ages. If the global fertility rate reaches 2.1 during this century, the population is projected to reach about 10 billion by 2100 (Withgott and Laposata 2015). If the TFR remains around 2.6, the population will be 16 billion and still increasing by 2100 (Withgott and Laposata 2015). However, if the fertility rate drops below replacement, to 1.6 children, the population will peak at around 8 billion in 2050 and decrease to about 6 billion by 2100 (Withgott and Laposata 2015). Fertility rates often depend on social and cultural factors, and not just resource availability. There are also social and economic reasons to encourage more population growth, as you will learn during the video shown in this lab.

Does the human population have a carrying capacity on the Earth, and if so, what is it? Humans depend on natural resources, and some might be limiting factors in the near future. Earlier predictions of food and water shortages have not come true at a global level, and some researchers argue that human ingenuity will continue to find new and improved ways to produce food and treat water. Other researchers suggest that there are biological limits to critical resources, and once those limits are reached, the birth and death rates might change rapidly. Questions about human carrying capacity also reflect individual ecological footprints and the choices about what type of environment people would like to live in. If everyone wanted to live like people in the United States, with large ecological footprints, then the carrying capacity would likely be smaller than our current population. If everyone chose to live more like people in Bangladesh, with a small ecological footprint, then the carrying capacity might be larger than the current population size. Some of the complexities of human population growth and consumption are explored in the video you will watch in this lab, The People Paradox, part of the World in the Balance series from NOVA.

Activities

1. *Human population growth during lab period.*
   Write down the global and US population sizes at the beginning of lab using http://www.census.gov/popclock/. Check again at the end of lab to determine how much those populations changed during the period.

   Start time: _____

   End time: _____

	Starting size	Ending size	Change
Global			
United States			

2. *The People Paradox video.*
   This video looks at the costs and benefits of population growth, focused in three contrasting regions: India, Japan, and Kenya. Look for environmental, economic, and social issues related to population growth in each country.

3. *Doubling-time calculations*
   Each group will be assigned five countries from different regions around the world by the TA. The 10-year growth rates for all the countries are listed on the Double Up activity handout, available on Blackboard. Use Excel to calculate how long it will take each country to double from 50 individuals to at least 100. Note that the ending population might be well over 100. There is an example file posted on Blackboard if you need help using Excel.
   After the calculations are finished, use Excel to graph the population growth for all five countries on a single plot. See Fig 3.2 for an example.

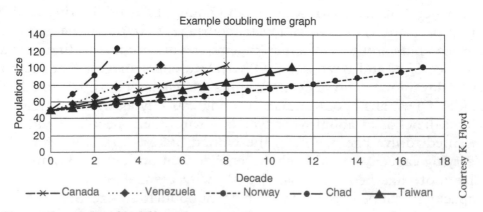

Figure 3.2. Example of the doubling time graph.

4. *Population pyramids.*

   Population pyramids, also called age structure diagrams, show the proportion of the population in different age categories for males and females. These diagrams are useful when predicting future population growth because they show what proportion of the population is currently of reproductive age, and what proportion will become reproductive in the next 15 years. Populations with many pre-reproductive individuals will likely experience continued population growth as those individuals start to reproduce, while populations with fewer such individuals will likely not increase. You can read more about population pyramids in the Human Population chapter of your textbook.

Your group will plot two population pyramids for each of the five countries, one showing the age structure in 1990 and one showing it in 2015. Download the data from http://www. census.gov/population/international/data/idb/informationGateway.php. Choose "Population by Five Year Age Groups" from the "Select Report" dropdown menu, find the correct country and year, then click submit. Scroll down the results page until you find the "Download all

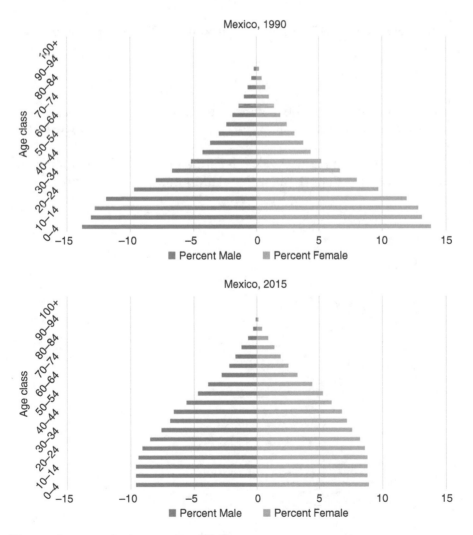

Figure 3.3. Example population pyramid figure.

Source: Data from Census Bureau

tables" link and choose the Excel version. To make the age pyramid, you first need to make the "Percent Male" values negative. Insert a column to the right of the existing "Percent Male" column, and use a formula to convert the numbers to negative values (type "=," then "-," the click the cell with the Percent Male value, then hit enter). You can copy the formula down to have Excel automatically perform the calculations. Select the cells for Age, the new negative Percent Male, and Percent Female (not including the first row of data that shows the total population across all age classes), and then insert a horizontal stacked bar graph. Give it a title and label the two data categories (See Fig 3.3). Repeat with the data from the other year, and copy and paste the two graphs next to each other in a Word document. Repeat for each of the countries in your group.

Assessments

Copy and paste the graphs (11 total) into a Word document. Print out one copy per group to turn in. Complete the rest of the questions in the assignment posted on Blackboard individually and turn in next week.

References

Withgott, J., and M. Laposata. 2015. *Essential Environment: The Science behind the Stories*. 5th ed. San Francisco: Pearson.

Resources

The NOVA website for the World in the Balance videos has additional videos and activities that can help further your understanding of this complex topic: http://www.pbs.org/wgbh/nova/worldbalance/

The Population Reference Bureau has a good explanation of human population growth: http://www.prb.org/Publications/Lesson-Plans/HumanPopulation/PopulationGrowth.aspx

Toa55/Shutterstock.com

Algal Biofuels and Population Growth

Learning Objectives

Part 1 (first week):
1. Describe potential environmental benefits of biofuels
2. Apply principles of the scientific method to design an experiment
3. Distinguish between dependent, independent, and control variables
4. Explain the importance of nitrate and phosphate to growth of primary producers (e.g., algae and plants)

Part 2 (second week):
5. Use Excel for basic calculations and constructing graphs
6. Interpret graphs
7. Describe trends in data
8. Distinguish between density-independent and density-dependent population growth
9. Relate resource (nutrients) availability to population growth
10. Identify sources of error and ways to reduce that error in future experiments
11. Use data to make recommendations for potential biofuel businesses

Importance

Transportation relies heavily on the use of petroleum, a fossil fuel also called oil. Extracting petroleum has negative environmental consequences, and burning petroleum (as gasoline or diesel) creates air pollution that can contribute to climate change. The United States has high demand for petroleum. Alternative fuels with a lower environmental impact are needed to shift transportation away from petroleum. One promising alternative is biofuels: ethanol or biodiesel. Ethanol is produced by fermentation of plant sugars, while biodiesel can be made from vegetable or algal oils or animal fats. Because biofuels are generated from organisms that were recently harvested, they are considered to be renewable energy and potentially have fewer negative impacts on the environment than fossil fuels.

Large amounts of plants, algae, or animals are needed to produce enough biofuel to meet current transportation demands. Algae can have very high rates of growth, and have the potential to be the most sustainable source of oils for the production of biodiesel. The rate of population growth for all organisms depends on the availability of resources. However, not all resources are equally important for growth, and a resource that is critical to survival and reproduction, and is relatively rare, will have a greater impact on how fast a population grows.

Cost-effective production of algal biodiesel requires an understanding of how to manipulate resource availability to maximize population growth rates.

Experiments are important components of scientific research, but are incomplete without collecting, analyzing, and interpreting data. Once analyses are complete, it is necessary to review how the experiment was conducted. In all cases, researchers should critically evaluate their experimental design and think about recommendations for future experiments. They also need to interpret their results in the context of the goals of the experiment: support for the hypothesis, answer to the research question, and guidance for applications (new policies, improved production of crops, etc.). In this lab we will conduct an experiment to determine how limiting resources (nitrate and phosphate) impact algal growth and, consequently, biofuel production. We will then analyze, interpret, and discuss our results.

Guiding Questions

How does the design of an experiment facilitate testing hypotheses and making predictions?

How feasible is it to produce transportation fuels that are both renewable and have lower environmental impacts?

Is nitrate or phosphate a more important resource for algal population growth?

Will humans change our population growth rates in anticipation of reduced resource availability, or are we more likely to experience population crashes when resources become limiting?

Vocabulary

1. *Microalgae*—Microscope unicellular algae; and there are thousands of species.
2. *Petroleum*—A liquid fossil fuel formed from partially decomposed organic material (mostly algae) that was buried under sedimentary rocks and subjected to high heat and pressure, converting it to a liquid over hundreds of millions of years.
3. *Feedstock*—In terms of biofuels, the raw material used to make the fuel (plants, algae, animal fat).
4. *Biofuels*—Energy sources made from living organisms or the waste they produce. The most common liquid biofuels are ethanol and biodiesel.
5. *Primary producers*—Organisms like algae and plants that can convert sunlight and CO_2 into sugars by the process of photosynthesis.
6. *Photosynthesis*—Biochemical process that uses light energy to convert CO_2 into sugars, primarily glucose. The type of photosynthesis performed by plants and microalgae also uses water and produces O_2.
7. *Population growth rate*—The rate at which the number of individuals in a population change over a given time period. The growth rate is a function of the birth and death rates and immigration and emigration rates.
8. *Density-independent population growth*—The growth rate does not depend on the size of the population. Also called exponential growth.
9. *Density-dependent population growth*—The growth rate depends on the size of the population. As the population size increases, the growth rate generally decreases as important resources become more limiting. Also called logistic growth.

10. *Limiting resource*—The resource that determines how the population grows, usually the scarcest resource.
11. *Growth medium*—A liquid or gel designed to support the growth of microorganisms. This experiment uses a growth medium developed in the Marine Biological Laboratory in Woods Hole, MA, that has a composition similar to freshwater, and is abbreviated MBL.
12. *Supernatant*—The liquid lying above the material deposited by settling, precipitation, or centrifugation.
13. *Composite sample*—A sample that is made by mixing two or more individual samples.
14. *Independent variable*—The variable that is manipulated in an experiment, levels of the independent variable corresponds to experimental treatments. Also called the experimental variable.
15. *Dependent variable(s)*—The variable that is expected to change in response to the level of the independent variable. Generally the dependent variable is measured to determine the results of the experiment. Also called the response variable.
16. *Control variable(s)*—The variables that should remain constant across experimental treatments. These variables could impact the dependent variable, so must be identical across treatments to try to isolate the effects of the independent variable to the extent possible.
17. *Average (mean)*—A measurement of the middle of a set of numbers, calculated by summing a list of numbers and dividing by the total number of numbers.
18. *Range*—The span between the lowest and highest values.
19. *Standard deviation*—A measure of the variation or dispersion in a list of numbers. The standard deviation gives the average distance that a given number is from the mean. Large standard deviations reflect higher amounts of variation, while small standard deviations reflect smaller amounts of variation in the data.

Introduction

Part 1: Transportation and biofuels

Transportation uses about 30% of all the energy consumed in the United States, and about 90% of that energy comes from petroleum, a fossil fuel (Davis et al. 2015). The United States consumes almost 20 million barrels of petroleum per day (840 million gallons), 21% of the world total, and about half is imported from other countries (Davis et al. 2015). The extraction, processing, transportation, and combustion of petroleum have negative environmental impacts, including air pollution and oil spills.

Vehicles using petroleum products (gasoline or diesel) emit particulates, nitrogen and sulfur oxides (NO_x and SO_x, respectively), carbon monoxide (CO), and carbon dioxide (CO_2), among other pollutants. You can find more details about the particular environmental impacts of each in the Air Quality and Pollution Control chapter of your textbook (Clean Air Act section). Carbon dioxide is a greenhouse gas, and major contributor to human-caused climate change. Transportation contributes about one-third of the CO_2 emissions from fossil fuel consumption in the United States (Davis et al. 2015).

Given the negative environmental and health impacts associated with combusting petroleum products, we need to develop alternative sources of energy for transportation.

One option is to use liquid biofuels, such as ethanol or biodiesel. Ethanol is the most commonly used biofuel in the United States, mostly as a fuel additive to reduce air pollution (Davis et al. 2015), and the vast majority of it is produced by fermenting sugars in corn. Biodiesel can be produced from vegetable oil or animal fat. The production of crops and livestock has environmental costs associated with agricultural practices. You can read more about ethanol and biodiesel in the Renewable Energy chapter of your textbook. In addition, the oil yields of the commonly grown crops are relatively low, and replacing just half of the transport fuel needs of the United States with biofuels is estimated to require 200–600 million hectares (775,000–2.3 million square miles, Chisti 2007). This is more than the area currently used for growing crops in the United States (Chisti 2007)! Thus today's crops are unlikely to replace much of the petroleum used in the United States. However, one promising alternative source is algae.

Microalgae (hereafter "algae") are unicellular microscopic species that can be found in freshwater, brackish, or saltwater aquatic ecosystems. Just like land plants, algae photosynthesize using CO_2 and water (H_2O) and the energy of sunlight to produce sugars (glucose, $C_6H_{12}O_6$) and oxygen (O_2). Some of the sugars are then used to make fatty acids and oils. Under the right conditions, algae grow rapidly, doubling their biomass in a day, and can have high concentrations of oils required for biodiesel production (Chisti 2007). Some studies estimate that algae can produce oil yields 50X that of terrestrial plants, such as canola whose oil is commonly used for cooking (Chisti 2007). This would allow the production of enough algae to replace half of the US transport fuels with only 5 million ha, or 2.5% of the current area used for growing crops (Chisti 2007). Although most of the technologies required to grow and harvest algae, extract the oils, and produce biodiesel from the oils are feasible, the cost of biodiesel from algae is still too expensive to replace petroleum (Scott et al. 2010). The point when algal biodiesel is cost-competitive with petroleum depends on many factors, including the cost of the petroleum, the cost of growing the algae, and the cost of extracting the oil and processing it into biodiesel. One study found that the cost of producing 1 L of algal biodiesel was $2.80, while 1 L of diesel was only $0.49 (Chisti 2007). Research is needed to both improve the growth of the algae and to minimize the costs of the entire production pipeline.

In addition to carbon dioxide and sunlight, algae (and plants) require nutrients, including nitrogen and phosphorus. Nitrogen is used to synthesize proteins and nucleic acids (DNA and RNA), and phosphorus is used to synthesize nucleic acids, cell membranes, and adenosine triphosphate (ATP—the source of energy that drives chemical reactions in cells). Without sufficient quantities (and specific ratios) of both nutrients, the algae will not grow well. However, adding nutrients in algae production increases costs, so determining the minimum concentrations of each nutrient that maximizes algae growth will help to lower the cost of producing algal biodiesel.

The research question that we will ask with this lab is: How do nitrate and phosphate concentrations affect the growth of algae? If algal biofuels are to become viable and cost-competitive with petroleum, we need to determine the most efficient way to produce large amounts of algae. To answer this question we will measure how algae grow in synthetic freshwater with different concentrations of nitrate (NO_3^-, a source of nitrogen) and phosphate (PO_4^{2-}, a source of phosphorus). In addition, we will use algae as a model system to understand population growth in general. Our algae cultures will be closed systems with regard to the amounts of nutrients. Some questions to consider during this experiment are: Will the populations grow exponentially or logistically? How will the initial concentrations of nutrients affect the growth patterns? Are there comparisons between how these populations of algae grow and how the human population grows?

Part 2: Data collection, analysis, and interpretation

This is the most exciting part of most experiments. You will finish collecting data on algal growth and changes in nutrient concentrations, and compare how the starting concentrations of nutrients impacted overall growth rates of algae. You will also relate changes in nutrient concentrations to changes in growth rates. As algae use nutrients to grow, what happens to the concentrations of those nutrients, and how does that feed back to influence future growth rates? You will also need to think about the broader implications of the experiment. How can you use these results to help improve the yield of algae for making biodiesel? Remember that science is an *iterative process*. Think about your research question and hypothesis. Do the results of this experiment support your hypothesis? If they do not, what could be a reason? Was there a problem with the experimental design itself, or was your hypothesis misguided? How could you change the experimental design in the future to better address your question? It is extremely rare that only one experiment is done when trying to answer a research question. What would be the next step in trying to understand how resources impact algal population growth? What other experiments might you try?

We will examine how algal populations changed over time by plotting the absorbance of light (an estimate of population density) for each treatment on the same graph. This plot will help illustrate how nutrients affect population growth. We will also calculate the rate of population growth. In general, the growth rate of a population is determined by the balance of births versus deaths and immigration versus emigration. If there are more births and immigrants than deaths and emigrants, the population will grow. In our closed bioreactors, there is no migration, so the growth rates are entirely determined by births and deaths. Both birth and death rates are affected by the availability of resources in density-dependent growth. We will calculate the growth rate of the algae during each of the time periods between samples using the formula:

$$\text{Growth rate } (\mu) = ln\left(\frac{\text{Absorbance}_{end}}{\text{Absorbance}_{start}}\right) \div (\text{time}_{end} - \text{time}_{start}),$$

where the subscripts refer to the start and end of the particular time period and *ln* is the natural logarithm.

For example, if the results for the experiment are:

Day	0	2	4	7	
Absorbance	0.05	0.3	0.5	0.55	
Time period		1	2	3	

we will calculate the growth rates for three time periods (day 0 to day 2, day 2 to day 4, and day 4 to day 7).

The growth rate for the first time period will be: $ln\,(0.30/0.05) \div (2-0) = 0.90$.

What will it be for the second time period?

_____

For the third time period?

_____

At the *carrying capacity* the population effectively stops growing. As an example for the algae cultures, this might be an absorbance of 0.50 both on day 4 and on day 7. What would the growth rate be for that time period?

_____

When the death rate is greater than the birth rate a population will decline. For example, this might be an absorbance of 0.50 on day 4 and 0.40 on day 7. What would the growth rate be for that time period?

_____

In general, what will the growth rate be when the population is growing (positive, negative, or zero)?

_____

When it is at its carrying capacity?

_____

When it is decreasing?

_____

Calculations and Excel

As much as we try to control the conditions in an experiment, there is always some degree of random error that occurs. In our experiment, possible causes of random variation include slight differences in the subsample of algae that was initially added to each replicate, the location of the bioreactors relative to each other and the light source, and the amount of aeration. Can you think of other factors that might contribute to this variability?

_____

We include several replicates of each treatment to improve our confidence in the conclusions we can draw from the results. If a particular treatment impacts algal population growth, then we should see a consistent trend across all replicates. If there is no consistency in the results, perhaps that treatment does not affect population growth or there may be problems with the experimental procedures.

One challenge in using replicated treatments is understanding and interpreting the results. To help we can calculate summary statistics of the data. There is more information on summary statistics and the reasons for including replicates in Appendix 2: Summary S-Statistics and Excel. We will be calculating the mean and standard deviation of the replicates for each treatment as a way to summarize the results. The mean gives information about the location of most of the data, while the standard deviation gives information about the amount of variation in the data (are the values of each replicate similar or different?).

We will use Excel to calculate the means and standard deviations of the light absorbance, the growth rates, and the weight of the algal pellets for each treatment. We will also use Excel to make scatter plots of how the light absorbance, growth rates, nitrate concentrations, and phosphate concentrations changed over the course of the experiment. Each plot will include the data for all of the treatments, not just the one that your group tested. In addition, the plots for the light absorbance and growth rates will include the standard deviations as error bars. Refer to Appendix 2 for more detailed instructions.

Activities

Experimental setup

Basic design

You are going to make simple bioreactors (containers to grow cultures of living organisms) using 500 mL clear plastic bottles (preferably with straight walls), synthetic freshwater, and algae (*Chlorella vulgaris*). You will monitor the growth of the algae using a YSI photometer that measures the amount of red light absorbed by the algae. As the algae population increases, the amount of light absorbed will also increase. At the end of the experiment, we will measure the total algal biomass. We will also collect a sample of the artificial freshwater at the beginning and the end of the experiment to determine how algae growth affects the concentration of nitrate and phosphate, two key nutrients for plant and algal growth.

Experimental variables

Algae growth is affected by the amount of light, space, temperature, and nutrients. Although all of these resources are important, we are going to focus on the role of nutrients in this experiment as they are often the resources that limit population growth. The particular nutrient concentrations that we will test are described in Table 4.1. Each group will be assigned one of these treatments for their experiment, and we will compile the results from each group at the end of the experiment.

Table 4.1: The four nutrient treatments to be tested.

Treatment	Nitrate	Phosphate
Control	Baseline	Baseline
High nitrate	4X Baseline	Baseline
High phosphate	Baseline	4X Baseline
High nitrate and phosphate	4X Baseline	4X Baseline

After you know which treatment your group will test, answer the questions about your research question, hypothesis, and experimental design on the lab assignment (posted on Blackboard):

Algae bioreactor set up, day 0

Procedure

1. Rinse the three 500 mL transparent bottles that your group provided with distilled water. Make sure that no trace of the original beverage is present as it might affect the growth of the algae. Remove any labels from the bottles.
2. Get the bottle of the appropriate artificial freshwater for your treatment. Note that the synthetic freshwater is also called the growth medium, and the particular recipe that we use was developed at the MBL in Woods Hole, MA, so it is called "MBL."
3. Each 500 mL water bottle will serve as a replicate for the treatment your group is testing. Label each bottle with: (1) the names of all the members of your group, (2) the treatment, (3) the replicate number, and (4) the name of the TA and the lab date and time. For example: "Claire, Joe, and Claudia. Control, replicate 1. Monday 10:30–1:20, TA: Javier."
4. Measure 300 mL of the appropriate artificial freshwater into each plastic bottle.
5. Measure 100 mL of the algae culture into each plastic bottle. *Make sure to swirl the flask with the algae before use to ensure that each 100 mL of algae culture has similar concentrations of algae.*
6. Mark the current water level on the bottle (400 mL total). You will need to add deionized (DI) water to the bottles to replace any water lost through evaporation during the experiment, and marking the current level makes this much easier.
7. Mix the algae and the growth media in the bottles, then pour 10 mL from each into a

© K. Floyd

different glass vial specifically for use in the photometer (you will need three vials total).

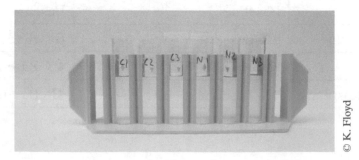

© K. Floyd

8. Measure the absorbance of each replicate using the **600 nm wavelength setting**. Record the results in the data table.
9. Return the sample to the correct bottle.
10. Add a section of air tubing and seal the top to the best of your ability with Parafilm.

© K. Floyd

11. Take the three bottles to the location on the back bench for your lab section. If there is not aluminum foil lining the benchtop and back wall, add those before continuing.
12. Hook up the three air tubes to a gang valve and connect the gang valve to an air pump. Plug the air pump into an outlet.

© K. Floyd

13. Adjust the air flow to each bottle so that there are bubbles, but not so many that it looks like the water is boiling. This aeration will help to keep the algae suspended in the medium and will constantly add CO_2 from the atmosphere to the algae culture.
14. Position the bottles so that light can reach all sides of all bottles.

15. Once all the groups have the algae cultures set up and the air flow adjusted, add aluminum foil to the front and sides to help reflect light to the cultures.

© K. Floyd

Nutrient monitoring, day 0

Procedure

1. Get a clean 50 mL centrifuge tube and label it with (1) the names of the people in your group, (2) the treatment, (3) the date and day of the experiment (Day 0 and 4) the lab section date and time.
2. Fill the centrifuge tube with 45 mL of the same growth medium that your group used to set up the algae bioreactors. You should have about 100 mL of growth medium remaining in the 1 L bottle after setting up all three replicates.
3. Place the tube in the centrifuge tube rack and your TA will store them in a refrigerator until the next lab, when you will measure the nitrate and phosphate concentrations from day 0 and day 7.

Days 2–5

Measuring algae growth during the week:

1. Your group is required to have at least two people go to the lab every two or three days to measure the concentration of the algae (see the timeline below, Table 4.2). *Failure to take the measurements during the week will result in a deduction of points from your lab assignment!*
2. Each time you measure the absorbance, first remove the air tubing and *fill the bottle up to the 400 mL line with DI water.*
3. Mix the solution so the algae are evenly distributed.
4. Add 10 mL into the glass vial and measure its absorbance at 600 nm. Record the value on the data sheet.
5. Repeat for each of the replicates.
6. Return the samples to the correct bottles and rinse the vials four times with DI water.

Table 4.2: Timeline for measuring algae population density

Lab meets:	First check	Second check	Final check
Monday (day 0)	Wednesday (day 2)	Friday (day 4)	Monday (day 7)
Tuesday (day 0)	Thursday (day 2)	Saturday (day 4)	Tuesday (day 7)
Wednesday (day 0)	Friday (day 2)	Monday (day 5)	Wednesday (day 7)
Thursday (day 0)	Saturday (day 2)	Tuesday (day 5)	Thursday (day 7)
Friday (day 0)	Monday (day 3)	Wednesday (day 5)	Friday (day 7)

Day 7

Overview

There are *three sets of data* that will be collected today: (1) the amount of light absorbed by algae (proportional to its population density), (2) the weight of the algae (proportional to its biomass), and (3) the concentrations of nitrate and phosphate (Fig. 4.1). Each group will collect these data for their particular treatment and then share those data with the rest of the class. Figure 4.2 summarizes the calculations and graphs required for each data type.

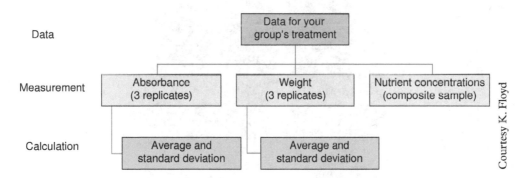

Figure 4.1. Flow chart of the data and calculations for the treatment your group tested.

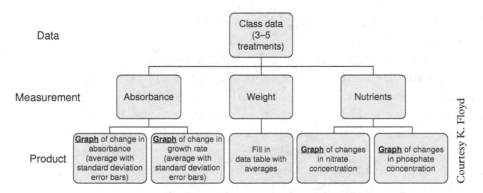

Figure 4.2. Flow chart of the calculations and graphs required for the class data (all of the treatments).

Data for light absorbance and weight will be summarized by calculating the average and standard deviation for the three replicates for each treatment. The light absorbance data will be used to calculate the growth of the algae during each time period. Finally, the changes in the light absorbance, the growth rate, and concentrations of nitrate and phosphate will be graphed. All calculations and graphs will be done using Excel.

Fill the algae bioreactors to the 400 mL line with DI water before doing any work today!

Weight of the algae

Procedure

1. Get three 250 mL centrifuge bottles. Make sure that each bottle and lid set have the *same number* on them, and that the *weight* of the bottle is written on the side.
2. Shake the algae bioreactor to suspend the algae throughout the culture. Measure 225 mL of each algae culture replicate into a centrifuge bottle. Write down the centrifuge bottle number that corresponds with each replicate on the data sheet.
3. Take the centrifuge bottles and the bioreactors to a balance. Add or remove the algae

© K. Floyd

solution to/from the centrifuge bottle until the mass is *exactly 300 g* (± 0.1 g) (*including the lid*).
4. Screw the lid on tightly, and give the bottles to your TA to centrifuge.
5. After the samples have been centrifuged, carefully pour the *supernatant* (the liquid) from each replicate into clean, labelled beakers for later use, being careful not to disturb the algal pellet. Use a transfer pipette to remove any remaining liquid. You want there to be as little liquid as possible without removing the algae pellet.
6. Weigh the bottles with the algal pellets (including the lids) and record the results on the data sheet. Subtract the weight of the empty bottle from the weight of the bottle plus algae to determine the weight of the algae itself.
7. Enter the data into the Excel file provided on Blackboard.
8. Calculate the average and standard deviation for your treatment (Appendix 2).

9. Divide the average mass of the algae pellet by 225 mL to determine the concentration of algae per mL of sample.
10. Share your data with the rest of the class.

Centrifuged sample showing the green algal pellet at the bottom and the clear supernatant. The arrows indicate the water line at the top of the supernatant.

Population growth

Procedure

1. Use the remaining 175 mL of algae in the bioreactors to measure the light absorbance in the same way as before.
2. Mix the algae culture in the bioreactor and add 10 mL to the glass tubes.
3. Measure the absorbance at 600 nm for each replicate and record the results on the data sheet.
4. Enter the data for all of the measurements (all four days) in the blank Excel template posted on Blackboard. If you did not record the data a particular day, enter "#N/A" in that cell.
5. Calculate the growth rate (μ) for each replicate for each of the three time periods. The formula is given in the introduction of this lab. It is easy to use Excel for these calculations. Be sure to enter the correct number of days for each time period (2 or 3, see Table 4.2).
6. Calculate the average and standard deviation for both the absorbance each day and the growth rate each time period.
7. Share your data with the rest of the class.
8. Once you have the data from the other groups for all the treatments you will make two graphs using Excel. The first graph will be the change in the light absorbance over the course of the experiment (Fig. 4.3). The second graph will be the change in the growth rates over the course of the experiment (Fig. 4.4). You will need to add error bars that show the standard deviation for each point, add axis labels, and a descriptive title for each graph.

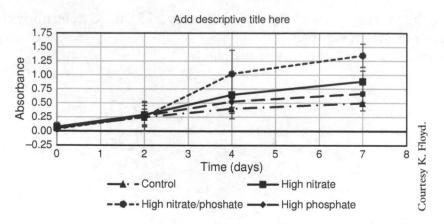

Figure 4.3. Example of Graph 1, changes in the light absorbance over time for all treatments. Note that the vertical error bars showing the standard deviation are included. You can manually select and delete the horizontal error bars that Excel automatically adds.

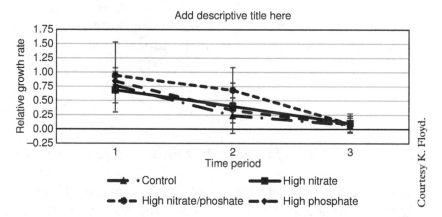

Figure 4.4. Example of Graph 2, changes in the growth rate over each time period for all treatments.

Nutrient concentrations

Procedure

You will need to measure nitrate and phosphate concentrations in the day 0 and day 7 samples. **Wear gloves at all times.**

Make a *composite sample* for day 7 by adding 15 mL from each replicate to a 50 mL graduated cylinder for a total of 45 mL of sample. Follow Part A if you do not have centrifuged samples and Part B if you do.

    A. If the sample has algae in it, filter the composite sample and use the filtrate (the clear portion) for nutrient testing. Connect the tubing to the vacuum line in the lab, and slowly turn on the vacuum just until the liquid begins to flow through the filter.

B. You do not need to filter the supernatant unless it is cloudy, which could happen if the algae pellet was disturbed when the supernatant was poured off.

Measuring nitrate (detailed instructions on the handout in lab):

a. The *day 0 samples for all treatments* need to be *diluted 1:20* before starting the procedure. Add 1 mL of sample to the plastic nitratest tube and fill to the 20 mL line with DI water. Our photometer can accurately measure nitrate concentrations up to about 5 mg/L, but our treatments range from about 20–80 mg/L, so the 1:20 dilution will bring the concentration of the sample into the measurable range.

b. The *day 7 samples* for the *control* and the *high phosphate* treatments *do not need to be diluted.* The samples for the *high nitrate* and the *high nitrate and phosphate* treatments should be *diluted 1:5* (4 mL of sample into the plastic nitratest tube, fill to the 20 mL line with DI water) before proceeding.

c. If the sample remains cloudy after adding the nitratest powder and tablet, shaking for 1 minute, and waiting about 5 minutes, you will need to filter the supernatant. Make a cone filter using the round filter paper and set it in a 50 mL beaker. Carefully pour the supernatant only into the filter and wait until all the liquid has filtered into the beaker. **Dispose of the filter in the waste bag, not in the garbage.**

d. Use the clear filtrate for the nitrate measurements.

e. Record the concentration (reading from the photometer) on the data sheet, and multiply by the dilution factor and by 4.4 to convert from N to NO_3^-.

Measuring phosphate (detailed instructions on the handout in lab)

a. The day 0 samples for the *high phosphate* and the *high nitrate and phosphate* treatments should be *diluted 1:1* (add 5 mL sample to the glass tube, fill to the 10 mL line with DI water) before proceeding. The *remaining treatments* and the *day 7 samples for all treatments do not need to be diluted.*

b. Record the concentration on the data sheet, and multiply by the dilution factor (if necessary).

Share the data with the rest of your classmates. Use Excel to make two graphs that show the change in (1) nitrate concentrations (Fig. 4.5) and (2) phosphate concentrations (similar to Fig. 4.5) from day 0 to day 7. Include all treatments on each graph.

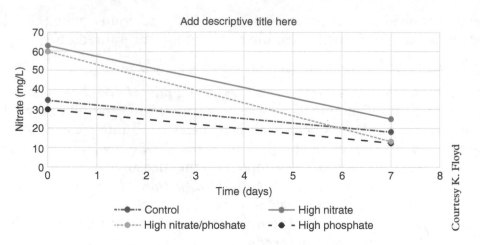

Figure 4.5. Example of Graph 3, changes in nitrate concentration during the experiment for all treatments. Note that there are no error bars present because we did not measure the nitrate in each replicate separately, so there is no information on the variation among replicates. The graph showing the changes in phosphate concentrations will look similar to this one.

Clean up:

   Any materials that have chemical waste on or in them need to be rinsed into the appropriate waste container for proper disposal.

1. Add 5 mL of bleach to each of the algae cultures. Fill the centrifuge bottle with the algae pellet with approximately 200 mL of water and 5 mL of bleach. Cap and shake the bottles to mix the bleach throughout. Let it sit for 5–10 minutes and then pour the solutions down the sink. Rinse the plastic water bottles to remove as much algae as possible and place in the box for recycling. Clean the centrifuge bottles as described below.
2. Remove the air tubing from the gang valves and rinse them with DI water. Wipe off any algae growing on the outside of the tubes. Place the tubing and gang valves in the appropriate plastic boxes.
3. Return the air pump to its box.
4. Remove the tape from the aluminum foil. Fold the aluminum foil that is in good shape for reuse. Aluminum foil in poor condition can be placed in the recycling box with the plastic bottles.
5. Refer to Appendix 3 for instructions on properly cleaning the glassware.

Assessments

Week 1 (setup): Complete the lab assignment posted on Blackboard, due at the beginning of lab the following week (week 2).

Week 2 (data collection, calculations, and graphs): *Copy each of the graphs into a single Word document to print* (four graphs total). Make sure that the graphs are not split between pages. Attach the printout to the lab assignment posted on Blackboard. All questions need to be answered individually, but you can discuss the questions with your group members before finalizing your answer.

References

Chisti, Y. 2007. "Biodiesel from microalgae." *Biotechnology Advances* 25 (3): 294–306.

Davis, S. C., Diegel, S. W., and Boundy, R. G, 2015. *Transportation Energy Data Book*. 34th ed. Oak Ridge, TN. http://cta.ornl.gov/data/index.shtml. Accessed May 9, 2016.

Stuart A Scott, Matthew P Davey, John S Dennis, Irmtraud Horst, Christopher J Howe, David J Lea-Smith, Alison G Smith. 2010. "Biodiesel from algae: challenges and prospects." Current Opinion in Biotechnology 21(3): 277–286. http://dx.doi.org/10.1016/j.copbio.2010.03.005.

Resources

TED talk by Jonathan Trent on growing algae for biofuels in floating pods. Interesting concept for using wastewater and other integrated methods to improve the economics and sustainability of the system. https://www.youtube.com/watch?v=X-HE4Hfa-OY, Accessed May 8, 2016.

Story about a large-scale algae biofuel plant in Alabama, also using wastewater. http://cleantechnica.com/2014/08/20/alabama-gets-first-world-carbon-negative-algae-biofuel/ Accessed May 8, 2016.

Information about the algae production facility outside of Columbus, NM. http://www.sapphireenergy.com/locations/green-crude-farm.html. Accessed May 8, 2016.

Introduction to Water

Learning objectives

1. Become familiar with the importance of freshwater to humans and all life
2. Learn about human impacts on freshwater availability and water quality
3. Understand challenges in meeting human demands for freshwater in arid environments
4. Learn about important chemical components of water and how to conduct water quality testing
5. Investigate water quality and aquatic life in a highly impacted urban river
6. Gain a basic understanding of wastewater, its treatment, and impacts of its release into freshwater systems

Importance

We take clean water for granted. Most of us turn on the tap with the assumption that the water will flow and be safe from chemical contaminants and potential pathogens. Throughout the United States, including in El Paso, drinking water is derived from the treatment of surface waters (predominantly lakes, rivers, streams) and groundwater. You may wonder: How do we assess the quality of these water sources? How are they treated to become safe for drinking? What other forms of life are found in surface waters and how does water quality impact them? The answers to these questions are often quite complex and require a variety of techniques and technologies to address them. The solvent property of water that makes it so special also makes it susceptible to pollution. Excess nutrients from human waste and agricultural practices enter into water systems (natural and man-made) as do many harmful chemicals from domestic and industrial uses. Maintaining adequate, high quality water for an ever-increasing human population while also supporting wildlife is one of the grand challenges facing our generation and those to come.

Introduction

Freshwater that could potentially be used for drinking and agricultural purposes comprises only 2.5% of all water on Earth. Of this only a small fraction is available as groundwater and surface water (0.03%); the remainder is locked in frozen ice caps and glaciers. Most of the water used by humans comes from surface waters. These waters are also home to a diverse array of fishes, invertebrates, and other organisms. Because of their connection to the landscape, surface waters are highly susceptible to pollution by substances that can harm the health of people and/or wildlife. Once water is polluted, it can either no longer be used or must be treated to make it usable. Most water pollution results from human activities such as agricultural runoff, wastewater and storm water return, industry, and mining. While there are laws and regulations in place to limit pollution of our freshwater resources, increasing human demands are stressing them. In the United States alone, we are extracting over 355 billion gallons of ground and surface water per day. In El Paso, we rely on water from three sources: surface water from the Rio Grande, groundwater, and desalinated water. The proportion of each source in the overall supply depends on time of year and the availability of surface water in the river.

Rivers not only supply water for human use and consumption but they also are living systems with diverse arrays of organisms that provide important ecosystem services. Specific components of water quality and flow rates determine which species will be found. Many species have tolerance limits for particular environmental conditions. These are factors that limit their ability to survive and/or reproduce. Some species are more sensitive to water quality than other, and the presence of these indicator species can provide a proxy for water quality. Managers often use the ratio of sensitive to tolerant taxa to assess water quality. Lastly, water quality and quantity are inextricably linked to the ecosystem services that can be provided. The world's rivers supply food to wildlife and humans across the globe. Other ecologically important services such as nutrient cycling, retention of nutrients and sediments in floodplains, recharge of groundwater, and decreased impacts of floods are also provided by surface waters. Groundwater filtration can act as a purification system for waters contaminated by certain compounds or microbes.

In the first lab of this unit you will learn about the importance of nutrients and other chemical components dissolved in water and how to measure them. You will be introduced to sources of water pollution, especially wastewater and how it is treated. The second lab is a field trip to the Rio Grande to give you an introduction to how environmental scientists monitor water quality, and to provide hands-on learning of some of the equipment and methods used to assess chemical, physical, and biotic components of water quality. You will observe how water quality is impacted by the influx of treated wastewater by sampling in the vicinity of a large capacity wastewater treatment plant. Plankton and macroinvertebrates collected during this field trip will be identified during the third lab in this series. You will look for certain species that are indicative of high water quality and learn how poor water quality has degraded habitat for fishes in the Rio Grande. Finally, you will learn and see how El Paso is using the process of desalinization to provide freshwater from our brackish groundwater reservoirs in order to meet consumer demands in our very water limited, desert environment.

© K. Floyd

Introduction to Water Quality

Learning Objectives

1. Describe major types of water pollution and their impacts
2. Identify national and local sources of water pollution
3. Gain proficiency with water quality parameters and how to measure them
4. Understand basic wastewater treatment methods and desired outcomes

Importance

Water is a resource fundamental to all living organisms. The particular chemical and physical characteristics of water influence the local biota and human uses. Water that becomes too polluted will have negative impacts on the biota, and will limit its use for humans. Many human activities pollute water, including industrial and wastewater discharge, agricultural and urban runoff, deforestation, and mining. Laws and local regulations attempt to limit the pollution entering local waterways, and in the United States, the water in most locations is less polluted now than it was in the mid-1900s. However, many water bodies are still polluted, and additional actions will be necessary to ensure clean and safe water for human and ecological uses.

Guiding Questions

What determines if water is safe to use?

What challenges does water pollution create for people? For wildlife?

Can clear water be polluted?

How do we obtain accurate measurements of water quality?

Vocabulary

1. *Pesticide*—A chemical used to kill unwanted organisms, including plants, insects, and fungi.
2. *Fertilizer*—A chemical that enhances the growth of plants and algae.
3. *Wastewater*—Everything that goes down a drain in building, including from toilets, showers, and sinks. Wastewater is treated to remove solids and microorganisms as well as some chemical compounds, and the treated liquid effluent is returned to the environment.

4. *Pathogen*—A bacteria, virus, protozoa, or fungi that causes illness.
5. *pH*—A measure of the acidity of a solution (the concentration of hydrogen ions, H^+). A pH of 7 is neutral, 0–7 is acidic, and 7–14 is basic. The units are on a logarithmic scale, so every change of one unit of pH represents a 10-fold change in H^+. Note that pH is unitless.
6. *Alkalinity*—A measure of the ability of a solution to neutralize an acid. The most common ions that contribute to alkalinity are carbonate and bicarbonate (the same material is common antacids).
7. *Dissolved oxygen (DO)*—Oxygen gas molecules that are dissolved in a solution. DO is critical to the survival of aquatic plants and animals. Some general concentration guidelines when considering test results are: 5–6 ppm (parts per million): Sufficient for most species; <3 ppm: stressful to most aquatic species; <2 ppm: fatal to most species.
8. *Hypoxic*—Solutions with low concentrations of DO, often defined as less than 2 ppm.
9. *Anoxic*—Solutions with no DO present.
10. *Ion*—An atom or molecule with an electric charge.

Introduction

The two primary concerns with water are **quantity** (how much) and **quality** (how usable). The El Paso Water Supply lab in this unit discusses how we obtain enough water to meet current demands, and this lab begins the exploration of water quality. Water pollution is the release of materials that can harm the health of people or organisms. Once water is polluted, it can either no longer be used (reducing the water supply) or must have additional treatments to make it usable.

Most water pollution results from human activities. *Agricultural runoff* can contain pesticides, fertilizers, and salts. *Wastewater*, including household waste from toilets, sinks, and showers as well as industrial and commercial waste, was discharged directly into water bodies for much of human history. Since the early 1900s in major U.S. cities, and spreading to most of the U.S. after the Clean Water Act of 1977, most wastewater is treated to remove suspended solids and to kill microbes prior to release of the effluent into the environment (Withgott and Laposata 2015). Untreated wastewater can still be discharged into the environment if a wastewater treatment plant fails, or if the community does not have sufficient resources to build a treatment plant. We will investigate the water quality of the wastewater effluent from the treatment plant for west El Paso, and how that effluent impacts the water quality of the Rio Grande in the next lab. You should review the process of wastewater treatment in the Water chapter of your textbook, or using the websites listed in the resources section of this lab. Pollution can also enter the water when rainwater *runoff from urban and industrial areas* washes gasoline, motor oil, or other chemicals into local waterbodies. Finally, both current and former *mining operations* can release toxic chemicals or acid drainage into receiving waters.

We will focus on five major categories of water pollution: microbiological, toxic chemicals, nutrients, biodegradable substances, and salts and sediments. You can read more about each, along with other types of pollution, in the Water chapter of the textbook. In addition, the chemical properties, aside from pollutants, of a water body can influence the types of organisms found there and what types of human uses are possible. You will collect water samples from different sites during the later field trip to the Rio Grande, and examine how the water chemistry relates to the biodiversity.

1. *Microbiological*

   Microbiological pollution focuses on disease-causing (pathogenic) organisms like bacteria, viruses, and protozoa. These can enter waterbodies from untreated human sewage discharge or runoff from feedlots, hog farms, or chicken farms (Fig. 5.1). People who then drink untreated contaminated water can become ill. Common water-borne diseases include cholera, dysentery, and giardiasis. Globally, there are around 842,000 deaths annually that are attributed to diseases associated with unsafe water supply (WHO 2014). Even in the United States there are an estimated 20 million cases of waterborne illness annually (Reynolds, Mena, and Gerba 2008). To reduce the infection risks, wastewater treatment plants disinfect the effluent using chlorine, ultraviolet light, or ozone prior to releasing it to the environment. Municipal water supplies also disinfect drinking water before distributing it to consumers. We will look for the presence of bacteria associated with fecal contamination at the Rio Grande using a *fecal coliform test*. A positive test indicates that fecal bacteria are present, but does not determine if the bacteria are pathogenic or not, nor if the bacteria originated from human or other animal waste. The main microbial pollution problem in the Rio Grande is *Escherichia coli*, a bacterium that normally inhabits the intestines of humans and animals. Most *E. coli* are harmless, but some are pathogenic, and can cause severe discomfort and diarrhea.

Figure 5.1. A cattle feedlot produces a lot of manure. If the manure enters into a local water body, it can substantially increase the number of bacteria and nutrient concentrations.

2. *Toxic chemicals*

   Toxic chemicals include pesticides, detergents, petroleum products, and industrial wastes. These can enter water bodies from surface runoff or direct discharge. These chemicals can be *acutely toxic* (causing death within two weeks) or *chronically toxic* (usually causing sublethal effects, such as growth, reproductive, or behavioral problems).

Some chemicals can accumulate in fishes and shellfish and poison people, and other animals that eat them. Others chemicals are acidic, and can lower the *pH* of the water body below the range that most organisms can tolerate. Most organisms require relatively neutral pH (approximately 5–9), and increased acid discharge can lower the pH to toxic levels. Some water bodies have high *alkalinity*, which acts to buffer the water against changes in the pH. Alkalinity is measured by the amount of carbonate and bicarbonate in the water, and both chemicals can neutralize acids. Waters in the eastern United States tend to have low alkalinity, and have suffered fish die-offs as a result of acid deposition (rain and snow that is more acidic than normal). In the western United States, the waters have higher alkalinity and are thus more resistant to pH changes. We will measure the pH and the alkalinity at the Rio Grande.

3. *Nutrients*

   Nitrogen and *phosphorus* are key nutrients for plant growth. You can read more about them in the Algae Biofuel lab, and we will also measure their concentrations in the soil treatments during the Food Production unit. Both nutrients can enter water bodies from agricultural runoff (crop fertilizer and animal manure, Fig. 5.1) and wastewater effluent. When these nutrients are added to lakes and rivers, they can cause algae "blooms," rapid increases in algal population density. The high algae densities can cover the surface of the water, blocking the light from reaching submerged plants and causing a nuisance to people swimming, fishing, or trying to treat the water for drinking (Fig. 5.2). Another problem begins when large numbers of algae die and sink to the bottom of the water body. The dead algae and submerged plants provide food for bacterial decomposers, and the bacteria populations boom. These bacteria consume the oxygen dissolved in the water as part of the decomposition process, and with their increased food source, they can consume the DO faster than it is being replaced. As the DO levels decrease, the water can become hypoxic or even anoxic. Larger animals either leave in search of oxic water or suffocate. As the populations of these animals decrease, the area is called a dead zone. There are more than 400 dead zones globally, including in the Gulf of Mexico where the Mississippi River enters and

Lodimup/Shutterstock.com

Figure 5.2. A water body covered by an algal bloom.

in the Chesapeake Bay (Diaz and Rosenberg 2008). The entire process of increased nutrients causing algae blooms which lead to decreased oxygen due to bacterial decomposition is called **eutrophication**. You can read more about eutrophication and dead zones in the Environmental Systems and Water chapters of the textbook. We will measure the concentrations of nitrate (NO_3^-) and phosphate (PO_4^{3-}) in the wastewater effluent and the Rio Grande.

4. *Biodegradable substances*

Many wastes are biodegradable, that is, they can be broken down and used as food by microorganisms like bacteria. Examples include human and animal wastes, paper products, and some types of industrial wastes. We tend to think of biodegradable wastes as being preferable to nonbiodegradable ones, because they will be broken down and not remain in the environment for very long times. However, as mentioned earlier the bacteria that perform the decomposition use oxygen as part of respiration. If a large amount of biodegradable wastes enter a water body, they can trigger a rapid increase in the bacterial populations, and a reduction of DO causing hypoxic or anoxic conditions. Once DO is depleted, other bacteria that do not need DO take over. While *aerobic* microorganisms, those which use DO, convert nitrogen, sulfur, and carbon compounds that are present in the wastewater into odorless and relatively harmless oxygenated forms like nitrates, sulfates, and carbonates, *anaerobic* microorganisms (those that do not use DO) produce toxic and smelly ammonia, sulfides (e.g., hydrogen sulfide that smells like rotten eggs), and flammable methane ("swamp gas"). In addition to bacterial populations causing decreases in DO, water with fewer photosynthetic plants or algae has lower DO. Warm and stagnant waters also tend to have lower concentrations of DO. Wastewater treatment plants often stir and aerate the wastewater to supply sufficient DO to maintain aerobic conditions. Ponds and small lakes, such as Ascarate Lake, will often have fountains that aerate the water to prevent anaerobic conditions and fish kills. We will measure the concentration of DO in the wastewater effluent and the Rio Grande.

5. *Salts and sediments*

Some chemicals are dissolved in water, meaning that the individual molecules or ions of the substance are mixed directly in between the water molecules. Common ions include sodium (Na^+), potassium (K^+), calcium (Ca^{2+}), chloride (Cl^-), nitrate (NO_3^-), and phosphate (PO_4^{3-}). Sodium, calcium, and chloride are often called salts, so the measurement of the dissolved solids is sometimes called the salinity of the water. Salts can affect the taste of drinking water, cause the buildup of deposits on faucets and in pipes, and impact the types of biota that can survive in the water body. Salts can enter a water body through wastewater effluent, urban runoff, and from agricultural runoff and drains. Agriculture in arid regions experiences high rates of evaporation of irrigation water. Salts from irrigation water are left in the soil, and the increasing salinization of soils reduces the ability of plants to grow. One method to reduce soil salinization commonly used in El Paso is to use additional water to flush the salts from the soils into drainage ditches, which then deliver the salty water to the Rio Grande. The hot summers in desert regions also cause high rates of evaporation from the water bodies. Again, the salts are left behind, so the more water that evaporates, the saltier the remaining water becomes. We will measure the *total dissolved solids* (TDS), which includes all of the ions present in the water. Related to TDS is *conductivity*, a measurement of how well the water conducts electricity. The more ions present, the better the water can conduct electricity.

Large particles do not dissolve in the water, but can remain suspended in the water column. These are called *suspended solids*. Suspended solids are often eroded soil sediments that wash into the water body (Fig. 5.3). Any activity that exposes the soil to wind and rain, such as clearcutting, agriculture, or construction, can lead to increased amounts of suspended solids in the water. The turbulence (mixing) in the water keeps the particles suspended. If the water is very turbulent, larger particles and even rocks can be transported. As the water velocity slows and become less turbulent, the larger particles settle to the bottom, and only the very small particles, like silts and clays, remain suspended. Suspended solids can block sunlight from reaching the bottom of the water (the substrate), preventing submerged plants from growing. Some suspended particles also have toxic materials bound to their surface, and this can lead to accumulation of toxins in the sediments of water bodies. We will measure the amount of suspended solids using a measurement of the *turbidity* of the water. Turbidity is the milky or muddy appearance of water that occurs when the small suspended particles scatter the light. We measure turbidity by lowering a *Secchi disk* into the water until the divisions between black and white sections on the disc can no longer be discerned. We record the distance that the disk was lowered as the Secchi depth, and it is inversely related to the turbidity. The more turbid the water is, the shorter the distance that those divisions can be discerned. If the water is very clear, then the Secchi disk can be lowered a long distance and divisions remain visible.

VictorN/Shutterstock.com

Figure 5.3. Eroded soil washing into the water body (on the left side of the picture), increasing the concentration of suspended solids and thus the turbidity.

Activities

Today we will learn how to measure water quality parameters: nitrate, phosphate, DO, alkalinity, pH, and TDS. We are going to use distilled water for many of the measurements. Distilled water has the majority of the dissolved salts present in most water removed. You will repeat these measurements in the next labs using samples from three locations: wastewater channel, the Rio Grande upstream of the effluent, and the Rio Grande downstream of where the effluent enters. The detailed instructions for how to use each instrument for measuring these parameters are provided on Blackboard and in the lab, and you should review Appendix 3: Water Chemistry for information about how to obtain accurate results.

For all of the measurements, record your results on the lab assignment (posted on Blackboard), and answer the questions. Although the results will be the same for all members of the group, your answers to the questions need to be completed individually.

A. *Nutrients*

Measure the concentrations of nitrate and phosphate in the samples from the Algae Biofuel lab using the instructions provided in lab. Our test kits can only measure a certain range of each chemical. Your TA will help you determine the correct dilution factors. So you will need to dilute some of the samples before adding any reagents. The suggested dilutions are given in the nutrient testing part of the Biofuel lab.

B. *Dissolved oxygen*

We will investigate the impact that temperature has on DO concentrations. Which temperature water do you expect to have more DO, cold or hot? _____

For the cold water sample:

1. Fill a glass 250 mL beaker with 150 mL of distilled water and place in the ice bath. Add a few pieces of ice to the beaker, and add a thermometer to measure the temperature of the water.
2. Wait until the temperature reaches about 5°C, record the temperature below, and then pour the water into a plastic bottle. The bottle should be around half full.
3. Close the bottle and shake it vigorously for one minute. This will introduce oxygen from the air in the bottle to the water.
4. Pour the shaken sample into the sample bottle in the DO test kit, filling the bottle all the way to the top. Do not leave any air at the top, as the oxygen in that air will affect your measurements.
5. Measure the concentration of DO following the instructions provided in the kit.

For the hot water sample:

1. Fill a glass 250 mL beaker with 150 mL of distilled water. Place it on a hot plate and add a thermometer.
2. Heat it to about 40°C –45°C, record the temperature, and pour the water into a plastic bottle.
3. Close the bottle and shake it vigorously for one minute. The air in the bottle will expand a bit because of the hot water, so be careful when opening it after shaking.
4. Measure the concentration of DO as above.

C. *Alkalinity*

1. Use a graduated cylinder to measure 1 L of distilled water into a bottle.
2. Take a sample and measure its alkalinity. Remember to add only a few drops of the titration reagent at a time, then swirl the sample to mix.
3. Add 0.5 g of baking soda (sodium bicarbonate) to the bottle of distilled water, and mix well to dissolve it in the water.

4. Measure the alkalinity of the baking soda solution. If the alkalinity is greater than 200 mg/L, you will need to refill the titrator to the 0 line and then continue to titrate the same solution. Once the color change occurs, add 200 to the number on the titrator to obtain the total alkalinity.

D. *pH*

1. Fill a 250 mL beaker with 150 mL of distilled water.
2. Use either a pH meter or the YSI probe to measure the pH of the water. Remove the cap, rinse the probe with distilled water, place it in the water sample, stir gently, and then wait for the pH value to stabilize.
3. After measuring one sample, rinse the probe with distilled water again before testing the next sample.
4. Add 50 mL of vinegar (a 5% acetic acid solution), mix, and measure the pH. Be sure to rinse the probe before and after measuring the pH.
5. After you are done with your measurements, store the pH probe in a beaker of tap water.

E. *Total dissolved solids*

1. Fill a 1 L beaker with 1 L of distilled water.
2. Rinse the YSI probe with DI water, then measure the TDS, conductivity, and salinity.
3. Add 2 g of table salt (NaCl), stir well to mix and dissolve, and then measure the same three parameters. The concentration should be 2 g/L, or 2 parts per thousand (‰).
4. Rinse the probe with DI water and store in a beaker of tap water.

Clean up

Refer to Appendix 3 for details on how to clean the glassware. Chemical waste must be disposed of in the appropriate waste container. The baking soda, vinegar, and salt solutions can be poured down the sink. Everything that was used needs to be cleaned with glassware soap, rinsed four times with tap water, and four times with DI water. You can leave the glassware in the plastic tub to dry.

Assessments

Fill out the tables in the lab assignment with the results of each measurement. Answer all of the questions individually.

References

Diaz, R. J., and Rutger R. 2008. "Spreading Dead Zones and Consequences for Marine Ecosystems." *Science* 321: 926–9.

Reynolds, K. A., Mena, K. D., and Gerba, C. P. 2008. "Risk of Waterborne Illness via Drinking Water in the United States." In *Reviews of Environmental Contamination and Toxicology*, 117–58. New York: Springer.

Withgott, J., and Laposata, M. 2015. *Essential Environment: The Science behind the Stories.* 5th ed. San Francisco: Pearson.

World Health Organization (WHO). 2014. *Preventing Diarrhea through Better Water, Sanitation and Hygiene: Exposures and Impacts in Low- and Middle-Income Countries.* Accessed July 13, 2016. http://apps.who.int/iris/bitstream/10665/150112/1/9789241564823_eng.pdf?ua=1/&ua=1.

Resources

Information on toxics in Rio Grande from TECQ:
www.tceq.texas.gov/waterquality/monitoring/riosum.html

Water pollution in the lower Rio Grande, near Laredo and Nuevo Leon:
www.npr.org/2013/10/22/239631549/despite-efforts-rio-grande-river-is-one-dirty-border

Video about dead zones from NOAA: www.nnvl.noaa.gov/MediaDetail2.php?MediaID=1062&MediaTypeID=3&ResourceID=104616

Information about wastewater treatment plants in El Paso: www.epwu.org/wastewater/wastewater_treatment.html

A short video about the wastewater treatment process from a water utility in Canada. It starts with what can and cannot be put down a drain, then finishes with how the process works. www.youtube.com/watch?v=oaXth88i7rk

© K. Floyd

Water Sampling Techniques

Learning Objectives

1. Investigate human impacts on local aquatic resources
2. Describe the process and purpose of treating wastewater
3. Practice water quality sampling techniques and procedures
4. Learn methods of collecting aquatic invertebrates found in the Rio Grande

Importance

Determining the health of freshwater ecosystems, including the Rio Grande, is of utmost importance. The Rio Grande provides drinking and irrigation water for human use as well as habitat for wildlife. It also receives water from many sources, including groundwater, rainfall, surface run-off, wastewater treatment plant effluent, and industrial discharges among others. In today's lab we will characterize water quality in an urban stretch of the Rio Grande. The sampling site was chosen because it is possible to monitor water quality of the Rio Grande upstream and downstream of a wastewater treatment plant. After the wastewater is processed, much of treated effluent (cleaned water) enters the Rio Grande through a canal. By sampling these three sites, we will be able to compare how water quality and flow rates in the river are impacted by the addition of treated wastewater.

Guiding Questions

How does water flow affect water quality?

What are major water sources for urban rivers and streams?

How does the input of treated wastewater affect water quantity? Water quality?

How do we obtain accurate measurements of water quality parameters?

What techniques are used to determine aquatic biodiversity?

Vocabulary

1. *Wastewater effluent*—Treated wastewater that is returned to the environment.
2. *Pathogen*—A bacteria, virus, protozoa, or fungi that causes illness.

3. *Periphyton*—An assemblage of primarily photosynthetic microbes (algae, cyanobacteria, and diatoms) and detritus that is found growing on rocks, woody debris and other surfaces lakes, streams, and rivers. Many macroinvertebrates, snails, tadpoles, and fishes are among the other aquatic life that feed on periphyton.

4. *Plankton net*—A net with a very fine mesh size that traps small aquatic organisms while allowing the water to flow through.

5. *Kick net*—A net with a more coarse mesh size that is attached to a handle. The net is placed on the river sediments and the users kicks rocks, sediments upstream of the net dislodging them so that they flow into the net.

6. *Flow rate*—The rate at which water that moves past a fixed point in a river during a given time interval (meters/second). A more accurate measurement would include calculating the volume of water that passes the point (cubic meters per second, cms).

7. *Secchi depth*—A measure of water clarity or turbidity. The depth is determined by lowering a disk with alternating black and white sections through the water column until the two colors cannot be discerned. The depth is then recorded.

8. *pH*—A measure of the acidity of a solution (the concentration of hydrogen ions, H^+). A pH of 7 is neutral, 0–7 is acidic, and 7–14 is basic. The units are on a logarithmic scale, so every change of one unit of pH represents a 10-fold change in H^+. Note that pH is unitless.

9. *Alkalinity* —A measure of the ability of a solution to neutralize an acid. The most common ions that contribute to alkalinity are carbonate and bicarbonate (the same material is common antacids).

10. *Dissolved oxygen (DO)*—Oxygen gas molecules that are dissolved in a solution. DO is critical to the survival of aquatic plants and animals. Some general concentration guidelines to considering test results are: 5–6 ppm (parts per million): Sufficient for most species; <3 ppm: stressful to most aquatic species; <2 ppm: fatal to most species.

Introduction

As we learned in the previous lab, fresh water quantity and quality are of great concern worldwide, especially in arid regions such as the desert southwestern US. Water quality is influenced by patterns of water flow (amount of water, rate of flow, seasonal, and daily fluctuations), interactions with sediments including inputs from upstream erosion, and the movement of gases and chemical compounds across the sediment/water/atmosphere interface. These factors influence the assemblage of life forms that will be found (plankton, plants, macroinvertebrates, fishes, etc.). And all interact to form functional aquatic ecosystems (Fig. 6.1). You learned about the importance of nutrients, microbes, and physical factors in determining water quality last week and next week you will learn more about the biota of the Rio Grande when you will examine the samples collected today. In today's lab we focus on the Rio Grande watershed and upstream and local inputs that impact our water quality, and learning the sampling techniques and protocols used by governmental agencies that are charged with monitoring water quality.

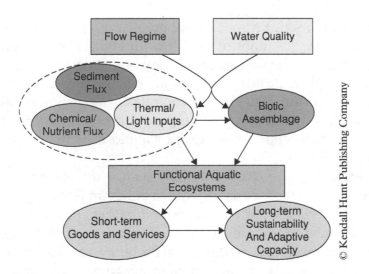

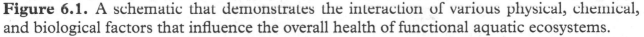

Figure 6.1. A schematic that demonstrates the interaction of various physical, chemical, and biological factors that influence the overall health of functional aquatic ecosystems.

Water quality monitoring

Because of the importance of having high quality water available for drinking and other household uses, there are many federal, state, and local agencies that are involved in monitoring water quality. In El Paso, these agencies include the El Paso Water Utilities, El Paso County Public Health, the Texas Clean Rivers program (a division of the International Boundary and Water Commission), the Texas Commission on Environmental Quality (TCEQ), the US Geological Survey (USGS), and the US Environmental Protection Agency (EPA). In addition, these agencies often have local partners such as UTEP and El Paso Community College. The type and frequency of monitoring varies among the agencies. For instance, El Paso Water Utilities monitors both regulated and unregulated compounds (see Resources section) as required by the EPA in drinking water that they produce and in the effluent that they release back into the environment after wastewater treatment. The EPA and US Food and Drug Administration set limits on amount of contaminants that can be present in drinking water (tap water, bottled water; see https://www.epa.gov/sites/production/files/2016-06/documents/npwdr_complete_table.pdf). They monitor six broad categories of water quality: disinfectants, disinfectant by-products, inorganic chemicals, organic chemicals, microorganisms, and radionuclides. This ensures that drinking water does not contain excessive salts, harmful substances or organisms, or other features that affect taste and/or odor. EPA also sets limits for maintaining water suitable for aquatic life such as plankton, fishes, birds, and mammals. Based on the best available science, they determine water quality criteria for both acute (short-term) and chronic exposure (long-term, life time, or multigenerational).

Another type of water quality monitoring is undertaken by the USGS. They have gaging stations located along the Rio Grande that measure temperature, pH, DO, conductivity, water depth, and discharge (rate and amount of water flow). Real-time data is available through their website (http://waterdata.usgs.gov/tx/nwis/current/?type=flow&group_key=basin_cd).

Finally, Rapid Bioassessment Protocols have been developed by the EPA to provide a fast, accurate, and consistent way to monitor health of the biotic community and to determine whether a stream or river can be designated for aquatic life use. In conducting a bioassessment, scientists follow explicit protocols to sample periphyton (primary producers growing on rocks and other substratum), macroinvertebrates, and fishes as well as assess habitat characteristics. Bioassessment is recognized as a way to integrate the cumulative impacts of multiple stressors (e.g., high salinity, high turbidity, and mercury contamination) over both short and long time frames.

Many of the potential water pollutants do not change the appearance or the smell of the water. The only way to determine the presence of the pollutant is through direct chemical testing or indirectly via bioassessments. It is important to realize that you can't determine water quality by appearances only. Regular monitoring is required to ensure human and ecosystem health.

Since determining water quality can be critical for human and wildlife health, the protocols that are used to collect water quality data are highly regulated and stipulated in great detail. The USGS instructions for field sampling are over 1500 pages long! Although each agency may have their own set of protocols for certain aspects of monitoring, they work together to make them as consistent as possible. They also require that individuals collecting the data use uniform datasheets and maintain a chain of custody for all samples collected. The chain custody is a form that documents who handled a sampled, how samples were labeled, how they were collected, and analyzed (see the Resources section for a link to a sample form). In today's lab, we will not be using chain of custody forms since our samples will not be used for regulating water quality. We, however, follow some best practices as we collect our samples. For instance, to avoid outside contamination of a sample, we will collect the sample in a clean bucket that has been rinsed 2–3 times in river water. We will also wear gloves to protect ourselves as well as the samples. You do not want to contaminate your nutrient samples with sediments, dust, or particles that you may get on your hands as you unload equipment and place it on the ground. We will repeat measurements that tend to be variable (flow rate) and the YSI multiparameter probe will be calibrated for accuracy prior to the field trip.

Wastewater treatment

All the water that flows into a drain inside a house or business after use is sent to a wastewater treatment plant. This includes water from toilets, sinks, and showers. Wastewater treatment plants are designed to remove the organic materials and then kill potential pathogens before returning the water to the environment. Most wastewater treatment plants have three treatment stages (Withgott and Laposta 2015). The primary stage uses large settling tanks or basins, where the majority of the suspended solids fall to the bottom, helping to separate the solids from the liquid. The liquid is sent to secondary treatment, where microbes like bacteria consume the remaining suspended particles of organic material. Secondary treatment ponds are often aerated to provide the bacteria sufficient concentrations of DO. Review the Water Quality lab for more details on DO and biodegradable substances. After the secondary treatment the water is very clear, but may still contain pathogens or potentially dangerous chemicals. Tertiary treatment disinfects the water using methods such as chlorination or ultraviolet light treatment. Some wastewater treatment plants will also take additional steps to remove nutrients like nitrate and phosphate, but this generally adds to the cost of the entire treatment process. Wastewater tends to be high in these nutrients because of the human waste, food scraps, and some types of soaps and detergents (EPA n.d.). After these final treatments the treated water, now called effluent, is returned to the environment. This often a river, lake, or ocean. In northeast El Paso, the Fred Hervey Water Reclamation Plant injects the effluent into the Hueco Bolson to help replenish the groundwater supply (EPWU, Fred Hervey). We will be sampling the effluent from the John T. Hickerson Water Reclamation Facility, which handles domestic and industrial wastewater from west El Paso and has a capacity to treat >17.5 million gallons per day (MGD) (EPWU, John Hickerson).

This process of returning treated wastewater effluent occurs along the entire Rio Grande. The Albuquerque Bernalillo County Water Utility discharges around 50 million gallons of effluent in the river each day (ABCWUA n.d.), while the Las Cruces Water Utility discharges around 10 million gallons daily (City of Las Cruces 2008). The quality of the water discharged by upstream users directly impacts the quality of the water available to downstream users. This is one reason that wastewater treatment and the water in the river are monitored by the agencies mentioned earlier.

Activities

Today we will measure water chemistry and physical parameters in three locations: upstream of wastewater treatment plant outflow, wastewater treatment plant outflow, and downstream of wastewater treatment plant outflow. The downstream sample should be collected where the clearer outflow mixes with the cloudier river water (see the picture at the beginning of the lab). The water quality parameters include those measured last week in lab: nitrate, phosphate, DO, alkalinity, pH, and total dissolved solids (TDS). In addition, we will measure the temperature, turbidity, and flow rate. We will also collect water samples to use for biodiversity monitoring next week and start fecal coliform tests. All of the water quality parameters will be measured during the field trip except for nitrate and phosphate. You will save filtered water samples to use for measuring those nutrients during next week's lab.

The class will be divided into three groups. Each group will measure the water quality and collect samples for a single location, and then share those data with the rest of the class. Although the data will be the same for the entire class, your answers to the lab assignment must be done individually.

Some people in each group will potentially be able to enter the water while wearing waders and a life vest to collect water samples. Read the River Safety instructions in Appendix 1 before entering the water. No one is required to enter the water if they do not choose to do so, and only people who can swim are allowed into the water. *Your safety is the primary concern during the field trip!*

Water chemistry

1. Use a bucket to collect a water sample from your assigned location. First rinse the bucket with sample water several times, and then collect the sample from just upstream of where you poured the rinse water. This prevents your sample from contained extra suspended solids stirred up by the rinsing process.
2. Take the bucket back to the area near the vans. Use the hand filter and filter about 100 mL of sample. If the water is very turbid (cloudy), start with a smaller volume because the filter is likely to clog. Go slowly and be careful not to squeeze the pump handle too hard as it can easily be broken.

3. Use the filtrate (the clear sample that passed through the filter) to measure the **DO** and **alkalinity** using the LaMotte test kits. The instructions should be in the test kit boxes.

4. Pour about 50 mL of filtrate into a clean and labeled Whirl-Pak to use for measuring nitrate and phosphate next week. The Whirl-Pak needs to be labeled with the sample location, the date, and the lab section. It should also be labeled as "filtered sample."

5. Work with the TA to use the YSI probe to measure the **temperature, pH, TDS** (and the related measures called **conductivity** and **salinity**), and **DO**. These measurements need to be taken directly in the sampling location, not using the sample in the bucket. Lower the probe into the water, but do not let it sink into the substrate. Record the values from the display once they stabilize.

6. Dispose of all water chemistry samples in the waste container, and rinse the glassware with distilled water. Make sure that all the glassware is returned to the correct kit.

Physical parameters

1. Measure the amount of suspended solids (the turbidity) using a Secchi disk.

   a. Start with the Secchi disk at the surface of the water, and slowly lower it down while watching from directly above.

   b. Continue lowering it until you cannot discern the difference between the black and white sections. That depth is called the Secchi depth, and is a common measure of turbidity.

   c. Note that the Secchi depth is *inversely related* to the amount of turbidity. If the water is very turbid, then you will not be able to lower the disk very far, and the Secchi depth will be small. If the water is very clear, then you will be able to lower the disk much more, and the Secchi depth will be larger.

2. Measure the flow rate by timing how long it takes a sample of dye to travel a preset distance.

   a. Measure 25 or 50 m along the location (longer for the downstream, shorter for the outflow).

   b. Dissolve a biodegradeable, non-toxic, fluorescent dye tablet in the bottle with some sample water. Wear gloves and be careful not to get any dye on your clothes. It does wash out eventually.

   c. Have one group member act as the time keeper. When they are ready, have a second student pour the dye into the water at the start of the measured distance as the time keeper starts the stopwatch (using the timer on their phone or watch).

   d. Walk along the bank following the dye. It tends to disperse once it is in the water and it can be difficult to see.

   e. Stop timing as soon as the dye reaches the end of the measured distance.

   f. Divide the distance traveled by the time to determine the flow rate. For example, if it took 100 seconds for the dye to travel 50 m, then the flow rate is 50 m ÷ 100 sec, or 0.5 m/s.

   g. If time allows, repeat the entire process two more times, and then calculate the average flow rate across all three replicates.

   h. Note that we cannot always measure the flow rate in the upstream location because of the vegetation growing on the banks.

Biotic samples

1. Fecal coliform bacteria are found in the intestines of warm blooded animals. They are used as an indicator of potential fecal contamination, but do not necessarily mean that the water contains pathogenic microbes or that the wastewater treatment process has failed.

   a. Obtain a fecal coliform test vial and label it with the sample location, today's date, and the lab section.
   b. Fill the vial to the line using *unfiltered* water. You can use the water in the bucket if there is any remaining.
   c. Do not shake the vial, and simply place it in the cooler to return to UTEP.
   d. The tablet in the vial will dissolve and form a gel with the nutrients that are specific to fecal coliform bacteria.
   e. Once you return to UTEP, place the vial upright in a small beaker and leave on a lab bench. After around 48 hours, the TA will place the vial in the refrigerator to slow bacterial growth until next week's lab.
   f. Next week, examine the sample. If there are any coliform bacteria present, they will grow and perform cellular respiration. That process produces carbon dioxide, which forms small bubbles in the gel. The carbon dioxide also reacts with water to form carbonic acid, lowering the pH of the gel. There is a pH indicator in the gel, so if the pH decreases the gel turns yellow. A positive test for fecal coliform bacteria is a yellow gel with bubbles, while a negative test is a red gel without bubbles.

2. Biotic sample of the water column, which is the water above the sediment, using a plankton net.

   a. Label a Whirl-pak with the sample location, today's date, and the lab section, and add "Plankton net sample."
   b. While holding onto the end of the rope with one hand (or having another student hold the rope), toss the net into the water as far as you can. Slowly pull it back to the shore, maintaining enough tension on the rope to prevent the net from dragging across the river bottom.
   c. Allow the water in the net to drain into the attached bottle. The plankton net has very small pores which act to trap any aquatic organisms present in the water column.
   d. Empty the sample into a Whirl-pak.
   e. Repeat 2–3 more times, until the Whirl-pak is about half full.
   f. Grab both ends of the Whirl-pak at the top and spin it several times (whirl it in circles) to close the top. Use the twist tie to ensure the sample won't leak out. There needs to be air trapped inside the top of the bag, otherwise most of the captured organisms will suffocate before we can examine them next week.
   g. If time allows, you can collect an additional sample, using a second Whirl-pak.
   h. Place the sample(s) in the cooler.

3. Biotic sample of the sediment using a kick net. A kick net has a D-shaped frame that allows it to be placed flush with the surface of the sediment.

   a. Label a Whirl-pak with the sample location, today's date, and the lab section, and add "Kick net sample."

    b. Place the kick net on the surface of the sediment downstream of where you are standing.

    c. Shuffle your feet or kick sediment into the net.

    d. Bring the net up, and transfer the sediment into the Whirl-pak. Add some water, and close as mentioned earlier.

    e. If time allows, you can collect an additional sample, using a second Whirl-pak.

    f. Place the sample in the cooler.

Clean up

1. Rinse the nets and waders in the outflow to remove as much sediment as possible. Lay them out to dry until everyone is ready to pack up the van.

2. All of the glassware in the water chemistry kits must be rinsed with DI water.

3. Rinse the bottle used for the flow rate dye in the outflow before placing it back in the supply box. Again, the dye tends to get everywhere if we do not rinse the container.

4. Pick up any trash around the site and place in a garbage bag to take back to UTEP. Let's leave the site cleaner than when we arrived.

5. Once back at UTEP, help the TA bring the equipment to the lab. Not everything needs to be returned to the lab each day, so ask your TA before taking equipment.

Assessments

Make sure that you have data from all three locations before leaving the lab. Answer the questions on the lab assignment individually.

References

Albuquerque Bernalillo County Water Utility Authority. n.d. *Wastewater Treatment Step 9 – Outfall.* Accessed November 11, 2016. www.abcwua.org/education/SWRP22_Outfall.html.

Baron, J. S., LeRoy Poff, N., Angermeier, P. L., Dahm, C. N., Gleick, , P. H., Hairston, Jr., N. G., Jackson, R. B., Johnston, C. A., Richter, B. D., and Steinman, A. D. 2003. "Sustaining Healthy Freshwater Ecosystems." *Issues in Ecology* 10: 1–18. http://www.esa.org/esa/wp-content/uploads/2013/03/issue10.pdf

City of Las Cruces Utilities Department. 2008. *Water and Wastewater System Master Plan Update.* Accessed November 11, 2016. www.las-cruces.org/~/media/lcpublicwebdev2/site%20documents/article%20documents/utilities/water%20resources/water%20and%20wastewater%20system%20master%20plan.ashx?la=en.

El Paso Water Utility (EPWU). n.d. *Fred Hervey Water Reclamation Plant.* Accessed November 11, 2016. http://www.epwu.org/wastewater/fred_hervey_reclamation.html.

EPWU. n.d. *John T. Hickerson Wastewater Treatment Plant, El Paso.* Accessed November 11, 2016. http://www.epwu.org/wastewater/wastewater_hickerson.html.

Environmental Protection Agency (EPA). n.d. *The Sources and Solutions: Wastewater.* Accessed November 11, 2016. https://www.epa.gov/nutrientpollution/sources-and-solutions-wastewater.

Withgott, J., and Laposata, M. 2015. *Essential Environment: The Science Behind the Stories*, 5th ed. San Francisco: Pearson.

Resources

The National Water-Quality Assessment (NAWQA) Program gives an overview of water-quality, trends in conditions these are affected by natural characteristics and anthropogenic activities. http://water.usgs.gov/nawqa/

The 1539 page US Geological Survey field manual for collection of water quality data can be found at water.usgs.gov/owq/FieldManual/

US EPA's Watershed Academy Web is can be found at www.epa.gov/watertrain

US EPA's Rapid Biological Assessment Protocols
cfpub.epa.gov/watertrain/pdf/modules/rapbioassess.pdf

Information regarding water quality in Rio Grande can be found at these sites:
The Texas Clean Rivers Program www.ibwc.state.gov/CRP/about.htm

Third phase of the binational study regarding the presence of toxic substances in the upper portion of the Rio Grande/Rio Bravo between the United States and Mexico.
ftp://63.96.218.8/RGToxicStudy.pdf

Resources from El Paso Water Utilities (EPWU):

How much do you know about your water?
http://www.epwu.org/whatsnew/420150729.html

A list of regulated parameters that are monitored by EPWU (English/Spanish)
http://www.epwu.org/water/pdf/dwr_2014.pdf

A link to a sample chain of custody form:https://alphalab.com/images/stories/M_images/madw_hc_coc.pdf

Introduction to Aquatic Biodiversity

Learning Objectives

1. Learn about local aquatic resources and biodiversity
2. Use both dissecting and compound microscopes competently
3. Identify organisms found in the Rio Grande
4. Describe components of biodiversity
5. Understand how water quality impacts aquatic life

Importance

Rivers not only supply water for human use and consumption but they also are living systems with diverse arrays of organisms. From the surface, they may seem lifeless but on closer examination an abundance of life forms are found: from the smallest organisms (bacteria, fungi, and protists) to small planktonic species (algae, rotifers, copepods, and cladocerans) to larger macroinvertebrates (mayflies, caddisflies, snails, crayfish, annelid worms—to name just a few) to still larger plants, amphibians, fishes, turtles, beavers, rats, ducks, and other birds. Specific components of water quality and flow rates determine which species will be found. Many species have tolerance limits for particular environmental conditions. These factors limit their ability to survive and/or reproduce. For example, many species cannot survive at low oxygen levels, while others may not be able to withstand high salt concentrations. In addition, nutrient concentrations and biotic interactions also play critical roles in determining the number of species present and their abundances. As you might expect, nutrients such as phosphate and nitrate drive algal (phytoplankton) abundance and in turn, the presence or absence of certain algal species can determine the composition of invertebrates that feed on them, and so on.

The world's rivers supply food to wildlife and humans across the globe. In addition, the presence of indicator species can provide a proxy for water quality. The larval stages of many insects are found in rivers and streams and are referred to as macroinvertebrates. Within these macroinvertebrates, some species are very tolerant of poor water quality (e.g., some midge larvae and some oligochaetes [relatives of earthworms]) while others are highly sensitive and can only survive in unpolluted waters with high oxygen concentrations. Managers often use the ratio of sensitive to tolerant taxa to assess water quality.

Guiding Questions

How does the aquatic community change in response to water quality?

What type of species can be used as indicators of water quality?

How can regional land use practices impact biodiversity in rivers and streams?

Would you eat a fish taken from the Rio Grande? Why or why not?

Vocabulary

1. *Wastewater*—Everything that goes down a drain in building, including from toilets, showers, and sinks. Wastewater is treated to remove solids and microorganisms, and the treated liquid effluent is returned to the environment.
2. *Pathogen*—A bacteria, virus, protozoa, or fungi that cause illness.
3. *Plankton*—Small aquatic organisms that are unable to swim against a current. They move passively in the direction of the water flow. Collected using a net with a very small mesh size (usually 64–500 microns).
4. *Macroinvertebrates*—Larger aquatic invertebrates that play various roles in the ecosystem (collecting, shredding, and processing large debris such as leaves; grazing on algae; preying on other macroinvertebrates and plankton). Collected using a kick net.
5. *Indicator species*—Species that have particular requirements for life (e.g., high oxygen levels, low amounts of pollutants, clear water).
6. *Food web*—The interconnections among organisms in a community based on who eats whom.
7. *Tolerance*—The ability of an organism to survive and/or reproduce under certain conditions.
8. *Abiotic*—Non-living; environmental factors such as temperature, pH, salinity, etc.
9. *Species richness*—Simply the number of species found in a given area.
10. *Species evenness*—The relative abundance of the different species present.
11. *Biodiversity*—The combination of species richness and evenness; the variety of life.
12. *Wet mount*—A type of slide preparation used to view living organisms under the microscope.

Introduction

The Rio Grande is not the river it once was. The river appears as a ribbon of life in a sea of the dry arid landscapes it flows through (Fig. 7.1A). The Rio Grande, known as the Rio Bravo in Mexico, is one of the most managed river systems in the Western United States. Like many river systems, the biological diversity in the Rio Grande has declined dramatically since the mid-1800s when the first of a long series of river management projects, including dams and rechannelization projects, were initiated. Human impacts have produced a river that flows intermittently and no longer supports a healthy and diverse aquatic ecosystem. This impact has been compounded the addition of excess nutrients and other pollutants introduced by humans (discussed further in the Introduction to Water Quality lab).

The Rio Grande was once home to 40 species of fishes (Calamusso, Rinne, and Edwards 2005). With the building of many dams along its course (e.g., New Mexico: Elephant Butte, Caballo; Texas: American Dam, Falcon), the water patterns and depth of the river were drastically altered. As you have likely noticed, in El Paso the river does not have much flow outside of the irrigation season (which is generally April–September) and is dry in many places for the majority of the year. Obviously, fishes and many other aquatic species cannot persist once the water dries. When water is released from the upstream dams, fishes and a variety of smaller species are released. You might be not be surprised to learn that most of the species that persist today (only 17!) are highly tolerant species (e.g., channel catfish, fathead minnow), and many are introduced species that did not occur naturally in the Rio Grande (e.g., mosquitofish, carp—Fig. 7.1B), and other are small fishes that recolonize quickly after release (e.g., minnows, shiners). Several fish species are at risk of extinction (gone from the face of the Earth) or extirpation (localized extinction). The Rio Grande Silvery Minnow is one of the most endangered species of freshwater fishes (Fig. 7.1C). When the river goes dry due to water diversion for human uses, the fishes are literally left to die in small, isolated pools. Sometimes heroic efforts are made to save the minnows.

Compared to many other large and important rivers, relatively little is known about the aquatic communities of the Rio Grande. This is especially true for the stretch of the Rio Grande in our area. Surveys are periodically taken by government agencies such as the International Boundary and Water Commission's Texas Clean Rivers Program, the Texas

© Vladimir Wrangel/Shutterstock.com. J. Lusk, USFWS, http://digitalmedia.fws.gov/cdm/singleitem/collection/natdiglib/id/17332/rec/1.

Figure 7.1. (A). Rio Grande and Bosque del Apache National Wildlife Refuge south of Socorro, New Mexico. (Map by K. Floyd with imagery from U.S. Geological Survey EROS Orthoimagery: https://raster.nationalmap.gov/arcgis/rest/services/Orthoimagery). (B). Common Carp, an introduced species of the Rio Grande. (C). Rio Grande Silvery Minnow, a native fish of the Rio Grande.

Commission on Environmental Quality, and US Geological Survey (IBWC 1994, 1998, 1999, 2004). In addition, there are scientific publications specific to certain taxa (fishes, turtles, etc.) but there is not a general species inventory repository.

As you have learned in the past two labs, assessing water quality is a complex process that involves measuring many chemical and physical aspects. In addition to flow, levels of salinity, oxygen, and nutrients can play a major role in structuring aquatic communities. Some aquatic organisms can tolerate low oxygen conditions for short periods but others cannot. Similarly, high levels of suspended particles (high turbidity) can clog the gills of some fishes, blocking their access to oxygen. High levels of salinity can increase energetic costs associated with osmoregulation in others. Being aware of these differences in tolerance, scientists often identify **indicator species** associated with particular types of habitats. For example, brown trout would be an indicator species signaling that water is high in oxygen. Some groups of macroinvertebrates can also be used as indicator species. Scientists and managers use the presence of EPT taxa (E̲phemeroptera, P̲lecoptera, and T̲richocoptera) to determine water quality (Fig. 7.2; EPA n.d.). These aquatic insect larvae need high oxygen levels and unpolluted water to persist and thrive. These abiotic factors act in concert with biotic interactions such as predator/prey interactions, competition, and parasitism. Figure 7.3 illustrates a typical riverine food web showing interactions among species. As you can see aquatic systems are complex and dynamic.

Figure 7.2. Examples of EPT taxa. (A). Ephemeroptera (mayfly) larva. Photo by E. Walsh. (B). Plecoptera (stonefly) larva molt. (C). Trichocoptera (caddisfly) larva.

Finally, why is determining biodiversity important? Long-term trends that can help us understand the health of an ecosystem. From what you read above regarding fish species diversity in the Rio Grande, what can you conclude about the condition of the river? How has the lack of flow and increased pollution changed how people use the Rio Grande for food and recreation, and how has it changed its aesthetic value? It is also important to measure diversity to obtain a sense of the complexity of the ecosystem. More complex ecosystems, those with more species and linkages, are generally more stable and resilient to future change. This is because several species may fulfill similar roles in the community and food web.

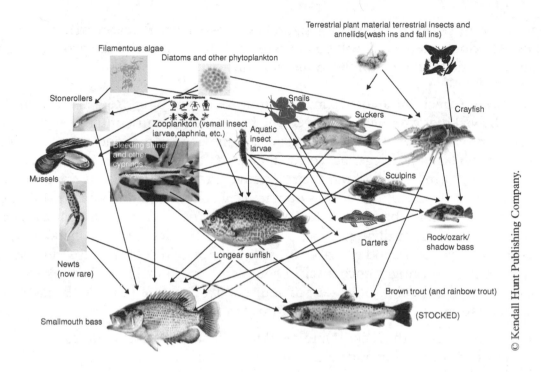

Figure 7.3. An example of an aquatic food web from the central United States. The algae and phytoplankton are the primary producers, which are eaten by primary consumers, which in turn are eaten by secondary consumers. If some species are extirpated from a region because of poor water quality, the impacts can ripple through the rest of the food web.

Activities

Overview

1. Watch videos showing diversity of micro-and macro-scopic animal and plant life in rivers
2. View organisms collected during field trip to the Rio Grande
3. Compare species diversity found at three sites in the river (upstream of wastewater treatment plant (WWTP), effluent channel, downstream of WWTP).
4. Analyze water samples from last week's lab
5. Check results of coliform bacteria cultures.

Today you will watch a series of videos to familiarize yourself with typical, small organisms found in riverine systems. Then you will observe species richness in samples from the three collection sites in the Rio Grande. You will also record the results of the coliform detection kit for each site.

Note: We did not specifically target larger organisms such as fishes and turtles since they are relatively rare and, importantly, special permission is needed to work with vertebrate animals such as fishes at the University as care is needed to minimize pain and suffering during handling and collection. This is spelled out in a number of Federal Laws including the Animal Welfare Act.

The detailed instructions for how to use microscopes (dissecting and compound) are provided on Blackboard and in the lab, and you should review Appendix 4: Using Microscopes.

Record your results on the lab assignment (posted on Blackboard), and answer the questions. Although the results will be the same for all members of the group, your answers to the questions need to be completed individually.

A. *Coliforms.* Check for a change in color from yellow to pink. Remember from last week that the vial that was filled with river water contained a nutrient tablet that is specific for coliform bacteria. As the bacterial grow and metabolize the food source, the pH of the solution will change triggering a color change in the media. Note: all mammals and birds release coliform bacteria from their digestive tracts. If your results are positive, it does not necessarily mean that there is human fecal contamination. What are other potential sources of coliform bacteria in the Rio Grande? *Hint:* Think about upstream land uses.

B. *Species richness.* Identifying species or taxonomic entities can be painstakingly difficult, even for experts. In today's lab, you will be provided with picture keys of common species that you might encounter in your samples. You will need to take pictures and categorize the species to the best of your ability. Feel free to use web-based resources as well (some are listed at the end of this lab).

1. Examine both the kicknet and plankton net samples collected from each site using both the dissecting and compound microscopes.
2. Gently mix the sample in the Whirl-Pak prior to removing a subsample.
3. Pour a small amount of the sample into a clean Petri dish for use with the dissecting microscope.
4. Use a disposable pipette to place one or two drops of sample onto a slide for use with the compound microscope. Refer to the instructions on how to prepare a wet mount slide in Appendix 4. If you observe an organism using the dissecting microscope, you can try to catch it with a pipette and transfer it to a wet mount slide.
5. Systematically scan the sample in the wet mount to find as many organisms as possible (Fig. 7.4).
6. Using the handout provided, attempt to identify the species you find at site, recording your observations on the lab assignment.
7. *If you do not observe any organisms in the first subsample that you examine, try taking another subsample and look for organisms.*
8. Summarize the species richness and relative abundance for each location in the lab assignment.

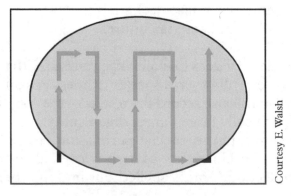

Courtesy E. Walsh

Figure 7.4. Schematic of how to systematically scan a wet mount sample.

C. Inspect your species richness results and compare with your water quality data for each site. Can you see any patterns? How could you test to see if these patterns are real?

Clean up

- Microscopes should always be left on low power and turned off.
- Only use *lens paper* to clean the microscope—never use paper towels or lab wipes. They will scratch the lens.
- Microscope slides and coverslips can be discarded into the glass waste container.
- Petri dishes should be washed, rinsed with distilled as water described in Appendix 3 and left to air dry.
- Plastic transfer pipettes should be thrown away.
- Samples should be returned to the cooler and your TA will take them to the incubator at the end of the laboratory.

Refer to Appendix 3 for details on how to clean the glassware. Chemical waste must be disposed of in the appropriate waste container. Everything that was used needs to be cleaned with glassware soap, rinsed four times with tap water, and four times with DI water. You can leave the glassware in the plastic tub to dry.

Assessments

Complete the lab assignment posted on Blackboard. Answer all of the questions individually.

References

Calamusso, B., Rinne, J.N., and Edwards, R.J. 2005. "Historic changes in the Rio Grande fish fauna: Status, threats, and management of native species." In *American Fisheries Society Symposium*. Vol. 45, 205. Bethesda, MD: American Fisheries Society.

EPA. n.d. *Rapid Bioassessment Protocols*. https://cfpub.epa.gov/watertrain/pdf/modules/rapbioassess. pdf, accessed on 7 Nov 2016.

IBWC. 1994. *Binational Study Regarding the Presence of Toxic Substances in the Rio Grande/Río Bravo and its Tributaries along the Boundary Portion between the United States and Mexico: Final Report by the International Boundary and Water Commission*, September, 250.

IBWC. 1998. *Second Phase of the Binational Study Regarding the Presence of Toxic Substances in the Rio Grande/Rio Bravo and its Tributaries along the Boundary Portion between the United States and Mexico (Final Report, Vol I of II, April 1998)* English Version, Accessed on November 7, 2016.

IBWC. 1999. *Second Phase of the Binational Study Regarding the Presence of Toxic Substances in the Rio Grande/Rio Bravo and its Tributaries along the Boundary Portion between the United States and Mexico (Final Report, Vol II of II, September 1999)* English Version, Accessed Nov 7, 2016.

IBWC. 2004. *Third Phase of the Binational Study Regarding the Presence of Toxic Substances in the Upper Portion of the Rio Grande/Rio Bravo Between the U.S. and Mexico (Final Report, June 2004)*. English Version, Accessed November 7, 2016.

Resources

Websites for identifying aquatic organisms:

    a. Texas Parks and Wildlife Department (TPWD). Common freshwater organisms in Texas rivers. Accessed on November 7, 2016. http://tpwd.texas.gov/education/resources/texas-junior-naturalists/bugs-bugs-bugs/common-freshwater-organisms.

    b. Images of microscopic life found in aquatic habitats. Accessed on November 7, 2016. www.funsci.com/fun3_en/guide/guide1/micro1_en.htm.

    c. EPT taxa: Accessed on November 7, 2016. www.cfb.unh.edu/StreamKey/html/biotic_indicators/indices/EPT.html.

Information about the fishes in the region: Lieb, C. S. 2000. Annotated Checklist of the Fishes of the Rio Grande Drainage, Dona Ana, El Paso, and Hudspeth Counties. Accessed on November 7, 2016. https://www.utep.edu/leb/pdf/fishlist.pdf.

There are many interesting YouTube videos of protists, freshwater aquatic invertebrates and vertebrates. Here are a few interesting ones:

    a. https://www.youtube.com/watch?v=9I8FP02y464, accessed on 7 Nov 2016.

    b. https://www.youtube.com/watch?v=tIMJWWpOrjw, accessed on 7 Nov 2016.

    c.https://www.youtube.com/watch?v=1HysvsXcmVI&list=PL8CKwQmut_uocdsabfPtCleZ-tzgysQn9g accessed on 7 Nov 2016.

Learn more about river ecosystems: http://sciencelearn.org.nz/Contexts/Toku-Awa-Koiora/Science-Ideas-and-Concepts/River-ecosystems, accessed on 7 Nov 2016.

HenryHo/Shutterstock.com

El Paso
Water Supply

Learning Objectives

1. Understand ways that water is used in the United States
2. Describe sources of El Paso's municipal water supply and how they change seasonally
3. Explain how to remove salts from water by the process of reverse osmosis
4. Learn about ways to reduce demand for water
5. Provide an assessment of the sustainability of water use in El Paso

Importance

Water is a critical resource for humans. In arid regions like El Paso, the demand for water can be greater than the sustainable supply. El Paso Water Utilities (EPWU) relies on three main sources for water: surface water from the Rio Grande, fresh groundwater from two local aquifers (Hueco and Mesilla Bolsons), and desalinated water (water with salts/minerals removed) produced from the brackish groundwater in the Hueco Bolson. Use of the fresh groundwater greatly outpaced its recharge in the 20th century in the El Paso region. This led EPWU to increase the use of surface water from the Rio Grande and to promote water conservation measures to reduce demand. However, drought conditions in the early part of the 2010s reduced the amount of available surface water, causing EPWU to resort to extracting more fresh groundwater to meet the demand. To help bolster freshwater supplies, EPWU starting desalinating brackish water from the Hueco Bolson in 2007. However, desalination requires more energy than treating surface or fresh groundwater, and is thus more expensive than the other freshwater sources and produces more greenhouse gases through energy combustion. As the population of El Paso and the surrounding region continues to grow, EPWU will either need to develop additional water supplies or continue to encourage customers conserve more water. In the long term, climate change is expected to cause increased frequency and severity of droughts in our region, reducing water flow in the Rio Grande, and further exacerbating water supply shortages. Finally, maintaining some level of minimum flow in the river is necessary to provide adequate water quality for wildlife, recharging groundwater, and for human recreation. Balancing all of the demands for water in the face of population growth and climate change is a complex challenge that will require contributions from scientists, policy-makers, and consumers.

Guiding Questions

How can we provide enough freshwater to supply an increasing human population?

What steps can individuals take to reduce water demand?

Should cities in arid regions like El Paso and Phoenix continue to increase in population in the face of decreasing water supplies?

What is the most important use of water: domestic, industrial, agricultural, or environmental? And how do we allocate a limited water supply among these uses?

Should water be appropriated to support wildlife? If so, what are the benefits of doing so?

Vocabulary

1. *Desalination*—The process of removing excess salts and minerals from water.
2. *Brackish*—Refers to water that has a salt concentration greater than freshwater but less than seawater.
3. *Total dissolved solids (TDS)*—A measurement of substances smaller than 2 microns that are dissolved in water. TDS includes substances such as nitrates, phosphates, calcium, sodium, and chloride. TDS is used to distinguish between freshwater, brackish water, and saline water.
4. *Public water supply*—Water treated and supplied for domestic, industrial, and commercial uses. Also called the municipal water supply.
5. *Water scarcity*—The point at which the water supply cannot meet the demands of all users, including for environmental uses such as habitat for wildlife.
6. *Potable water*—Water that is safe for human consumption.
7. *Environmental flow*—Amount of water that must flow in a river to maintain water quality and support ecosystems.
8. *Recharge rate*—The rate that water enters an aquifer, generally through percolation by rain and snow melt. Paving and development can reduce the recharge rate, and in many parts of the world the rate of pumping water out of an aquifer exceeds the recharge rate, leading to decreases in the water availability.

Introduction

Only 2.5% of all water on Earth is freshwater with salt concentrations low enough for drinking and irrigation of crops. Freshwater is often defined as <500 mg/L TDS, but the Texas Commission on Environmental Quality sets the drinking water standard at 1000 mg/L (TCEQ, 2015) and the El Paso Water Utility uses this criterion. Most freshwater (79%) is frozen in ice caps and glaciers, 20% is groundwater, and 1% is surface water. Groundwater is contained in subsurface aquifers, which are porous sections of rock, gravel, or sand. Surface waters include lakes, rivers, soil moisture, and atmospheric water vapor, and comprises only 0.03% of all the water on the planet. Most of the water used by humans comes from surface waters.

Global water trends

Human extraction of water increased almost 300% from 1950 to 2000 (UNEP 2008), and is projected to increase an additional 55% from 2000 to 2050 (OECD 2012). Around 700 million people currently suffer from water scarcity, and by 2025 that number is expected to increase to 1.8 billion people as a consequence of population growth and economic development (UNDESA 2014). Although it varies by region, globally most surface water (75%) is used for crop irrigation (UNEP 2008).

US water trends

The United States withdrew 355 billion gallons of ground and surface water per day in 2010 (Barber 2014). Water for cooling electric plants and irrigation were the two largest uses, and together with water for domestic uses accounted for 90% of all withdrawals (Barber 2014). In rural areas located far from public water supplies, groundwater supplies almost all of the water (Barber 2014). Groundwater withdrawals increase during periods of drought when surface water supply is limited.

El Paso water trends

EPWU supplies water from three sources: surface water from the Rio Grande, groundwater from the Mesilla and Hueco Bolsons (aquifers), and desalinated water from the Kay Bailey Hutchison (KBH) Desalination Plant, which treats brackish water found below the freshwater layer in the Hueco Bolson. The proportion of each source in the overall supply depends on time of year and the availability of surface water in the river (Fig. 8.1). Flow in the Rio Grande near El Paso relies on the release of stored water in Elephant Butte and Caballo Reservoirs, located nearly 200 km (125 miles) north of El Paso. Water is released primarily for irrigating cropland in southern New Mexico and the El Paso region. The irrigation season generally runs about 7 months, from March through September. During this period, EPWU can supply the city with their allotment of river water (water in the river is allocated to different users, including farmers and cities). During the rest of the year, or during droughts when there is reduced or no flow in the river, EPWU uses groundwater from the two aquifers and desalination to meet customer demand. This groundwater is extracted via wells.

Prior to the 1980s, groundwater from the Hueco Bolson provided most of the water supplied by EPWU (EPWU, Water Resources). Concerns about how long the Hueco Bolson could continue to provide freshwater of good quality led to several studies in the 1990s, which concluded that El Paso would deplete the Bolson in the mid-2020s and Ciudad Juarez would deplete its supply of groundwater by 2010 if the current rate of use continued (reviewed in Sheng et al., Fahy, Devere 2001). EPWU responded by increasing its use of surface water (building the Jonathan Rogers Water Treatment Plant in 1993) and promoting water conservation to reduce per capita demand (discussed later). Due in large part to the resulting reduction in groundwater withdrawals, a study in 2004 found that freshwater supplies in the Hueco Bolson would last at least 50 years (Hutchinson 2004). However, pumping still exceeds the recharge rate and water levels are dropping, but at a much slower rate than before.

The Hueco Bolson contains a large volume of brackish water (1000–3000 mg/L TDS) underneath a layer of freshwater. There is about six times as much brackish water as freshwater in the Bolson (Sheng et al., Fahy, Devere 2001), and this represents another potential source of potable water if salts are removed to meet drinking water standards. Some wells reach into the deeper brackish water, and deliver that water to the KBH Desalination Plant. The Plant opened in 2007, and uses reverse osmosis (RO) to produce up to 56 million L

(15 million gal) of freshwater daily (TWDB, 2014). The desalinated water is blended with freshwater from other wells before being sent for distribution, and the total capacity (desalinated plus blend water) is 100 million L (27.5 million gal) per day (TWDB, 2014). However, producing water via RO requires a lot of energy to push the water through the RO membranes, and the high energy use increases the operation costs. Desalinated water costs around $1.25–$2.60 per 3800 L (1000 gal), or about twice the cost required to produce freshwater from groundwater wells (Wythe 2014). Due to this higher cost, the desalination plant generally runs at minimum capacity unless there is increased demand, such as during the summer or drought conditions. The desalination plant has a recovery rate of 83%, meaning that for every 1000 L of brackish water treated, 830 L of desalinated water is produced (EPWU, Desalination Info). The remaining 17% is concentrate, water with higher salt content than the incoming brackish water. Disposal of concentrate is an important environmental issue associated with the desalination process. EPWU disposes its concentrate using deep well injection into the Earth about 35 km northeast of the plant (TWDB, 2014).

The total water supply that EPWU can deliver is about 130,000 acre-feet per year (4.2 billion gallons or 160 million m^3; EPWU, Water Resources). Under normal climatic conditions, about half of that supply comes from the Rio Grande, one third from the Hueco Bolson, one sixth from the Mesilla Bolson, and a small amount from the desalination plant (Fig. 8.1; EPWU, Water Resources). During a drought, such as 2013 (Fig. 8.1), the lack of water from the Rio Grande is compensated by increasing production from the other sources, with the majority coming from freshwater in the Hueco Bolson. Although EPWU included drought conditions when assessing the lifespan of the Hueco Bolson, climate change is expected to increase the length and severity of droughts in our region (Venkataraman et al. 2016). If severe droughts reduce the amount of water EPWU can withdraw from the Rio Grande in the future, other alternatives will be required, such as further reducing water use, importing water from farther away, or the planned construction of an advanced purification plant that will treat wastewater effluent to drinking water standards (www.epwu.org/water/purified_water.html).

So far we have discussed how the water supply is produced. The other side of the equation is the demand for water. If EPWU customers reduce the amount of water they use every day, then EPWU will not have to process as much water. There are many ways to reduce water demand. In agriculture, this could include using more efficient irrigation methods, or planting crops that are adapted to the local environment. For individual households, installing low-flow toilets and showerheads, reducing irrigation to landscapes or replacing lawns with native species, turning off faucets when not in use, and choosing a plant-based diet that requires much less water, can all help to reduce water use. EPWU and the El Paso city government have focused on water conservation since the 1990s. Some examples are rebates for replacing lawns with low water use alternatives ("xeriscaping"), restricted residential watering, giving away low-flow showerheads, and imposing fines for wasting water (Scott 2012). These efforts have helped reduce the per capita water demand from about 850 L per day (225 gallons) in the 1970s to about 490 L per day (130 gallons) in 2013 (Hutchinson 2004; EPWU 2014). Continued efforts to reduce demand and find additional water sources will be required as El Paso and Ciudad Juarez continue to grow.

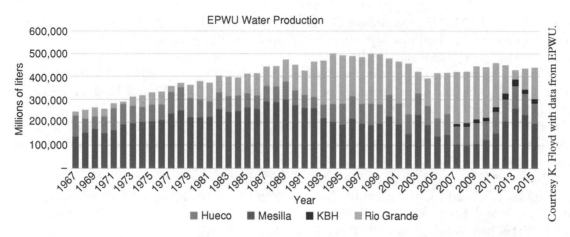

Figure 8.1. Total amount of water processed by the El Paso Water Utility (EPWU) and the source of that water. The Hueco and Mesilla Bolsons are groundwater, and the Rio Grande is surface water. The black bars labeled KBH are water produced by the Kay Bailey Hutchinson Desalination Plant.

Activities

You will visit the EPWU KBH Desalination Plant and the TecH$_2$O Education Center, both located off of Montana Ave. near the El Paso International Airport. At the Desalination Plant, you will have the opportunity to see equipment used in the desalination process and talk to plant operators about their careers. At the TecH$_2$O Center, you will explore sources of water, how water is treated to make it safe to consume, how treated wastewater effluent is used to reduce the demand for potable water, and ways to reduce your personal water footprint.

Assessments

Prior to coming to lab, complete the pre-lab quiz. The quiz will help you learn more about desalination in general and the EPWU desalination plant specifically, and will prepare you to ask questions at the plant.

You will need to print out the lab assignment available on Blackboard prior to coming to class. You will answer some of the questions while on the field trip and finish the rest as homework.

References

Barber, N. L. 2014. "Summary of Estimated Water Use in the United States in 2010." U.S. Geological Survey Fact Sheet 2014–3109, 2 p. Accessed May 8, 2016. http://pubs.usgs.gov/fs/2014/3109/pdf/fs2014-3109.pdf.

El Paso Water Utilities (EPWU). n.d. "Desalination Plant." Accessed May 8, 2016. www.epwu.org/water/desal_info.html.

EPWU. n.d. "Water Resources." Accessed May 8, 2016. www.epwu.org/water/water_resources.html.

EPWU. 2014. "El Paso Water Utilities 2014 Water Conservation Plan." Accessed May 8, 2016. www.epwu.org/conservation/pdf/Conservation_Plan_2014.pdf.

Hutchinson, W. R. 2004. "Hueco Bolson Groundwater Conditions and Management in the El Paso Area." El Paso Water Utilities Hydrogeology Report 04-01. Accessed May 8, 2016. www.epwu.org/water/Hueco_Bolson_Report.html.

Organization for Economic Co-operation and Development (OECD). 2012. OECD Environmental Outlook to 2050: The Consequences of Inaction. Paris: OECD Publishing. Accessed May 8, 2016. http://dx.doi.org/10.1787/9789264122246-en.

Scott, D. 2012. "Water Shortage Tests El Paso's Conservation Efforts." Accessed May 8, 2016. www.governing.com/topics/energy-env/gov-water-shortage-crisis-puts-el-pasos-conservation-efforts-to-the-test.html.

Sheng, Z., Fahy, M. P., and Devere, J. 2001. Management Strategies for the Hueco Bolson in the El Paso, Texas, USA, and Ciudad Juarez, Mexico, Region. *In* Bridging the Gap: Proceedings of the World Water and Environmental Resources Congress, Orlando, Florida, May 20–24. Accessed May 8, 2016. http://dx.doi.org/10.1061/40569(2001)346.

Texas Commission on Environmental Quality (TCEQ). 2015. "Rules and Regulations for Public Water Systems, Rule 30 TAC 290.271-275." Accessed May 8, 2016. www.tceq.state.tx.us/drinkingwater/pdw_rules.html.

Texas Water Development Board (TWDB). 2014. "Worth its Salt: El Paso Water Utilities Kay Bailey Hutchison Desalination Plant." Accessed May 8, 2016. www.twdb.texas.gov/innovativewater/desal/worthitssalt/doc/Worth_Its_Salt_Jan2014_KBH.pdf.

United Nations Department of Economic and Social Affairs (UNDESA). 2014. *Water Scarcity*. Accessed May 8, 2016. www.un.org/waterforlifedecade/scarcity.shtml.

United Nations Environment Programme (UNEP). 2008. *Vital Water Graphics—An Overview of the State of the World's Fresh and Marine Waters*. 2nd ed., Nairobi, Kenya: UNEP. Accessed May 8, 2016. www.unep.org/dewa/vitalwater/index.html.

Venkataraman, K., Tummuri, S., Medina, A., and Perry, J. 2016. "21st Century Drought Outlook for Major Climate Divisions of Texas Based on CMIP5 Multimodel Ensemble: Implications for Water Resource Management." *Journal of Hydrology* 534: 300–16.

Wythe, K. 2014. Everybody is Talking about it: Is Brackish Groundwater the Most Promising "New" Water? In *tx:H₂O Summer 2014*. Texas Water Resources Institute. Accessed May 8, 2016. http://twri.tamu.edu/publications/txh2o/summer-2014/everybody-is-talking-about-it/.

Resources

Short video from NASA about a study that finds increased probability of "megadroughts" in the western United States due to climate change. Accessed May 8, 2016. https://youtu.be/ToY4eeWsdLc.

A report from the UNEP about the state of the world's fresh and marine waters. Lots of interesting maps and graphics that show many different aspects of global water issues. Accessed May 8, 2016. www.unep.org/dewa/vitalwater/index.html.

A story from 2012 about El Paso's water supply and plans for the future. This period was during drought conditions. Accessed May 8, 2016. www.governing.com/topics/energy-env/gov-water-shortage-crisis-puts-el-pasos-conservation-efforts-to-the-test.html.

A news article that includes how Juarez and El Paso both use water from the Rio Grande and Hueco Bolson, and the challenges the increased growth of both cities puts on the water supply. Accessed May 8, 2016. https://journalism.berkeley.edu/projects/border/elpasodraining.html.

Introduction to Food Production

Learning Objectives

1. Describe current trends in food production and food security
2. Explain how current agricultural practices impact soil fertility
3. Summarize interactions between composting, soil fertility, and food production

Importance

Although food production has increased over the past 40 years, there are still almost 800 million undernourished people globally. This is actually a decrease in both the overall number and total percentage of undernourished people since 1990, even as the global population increased by almost 2 billion people during that same time. Food production must continue to increase over the next century, both to feed the people who are currently undernourished and to feed the 2–3 billion additional people that will be born during that time. However, many current agricultural practices require large inputs of energy, water, chemical fertilizers, and pesticides, and reduce soil fertility. Alternative agricultural methods that improve soil fertility and reduce the inputs will be needed, in addition to changes in food distribution and dietary choice, to sustainably feed 9–10 billion people.

Vocabulary

1. *Food security*—Having long-term access to adequate food for an active, healthy life.
2. *Undernourishment*—A measure of hunger; the proportion of the population whose food consumption is less than the number of kilocalories required to engage in low levels of activity.
3. *Soil fertility*—The ability of the soil to provide nutrients and water essential for plant growth.

Introduction

Hunger and food production

Ending hunger globally has been a goal of the United Nations (UN) for many years. In the fall of 2015, the UN announced 17 sustainable development goals (UN 2015a). Goal 2 is to "end hunger, achieve food security and improved nutrition, and promote sustainable agriculture" (UN 2015a). There are about 795 million people globally who are undernourished, 11% of the total population of over 7 billion people (Fig. 9.1; FAO 2015). The vast majority of those people (780 million) live in the developing world (FAO 2015). Although these are large numbers, there has actually been a substantial reduction in hunger over the past 30 years. In 1990, about 1 billion people were undernourished, and this was almost 19% of the global population of around 5.3 billion (Fig. 9.1; FAO 2015). The number of undernourished people today represents a decrease in overall percentage of the population of almost half because the population increased by about 2 billion people (Fig. 9.1). The current goal of reducing global hunger is facing multiple challenges, including continued population growth that is projected to reach more than 9.7 billion by 2050 and 11 billion by 2100 (UN 2015b), climate change, and environmental degradation of current croplands and pastures.

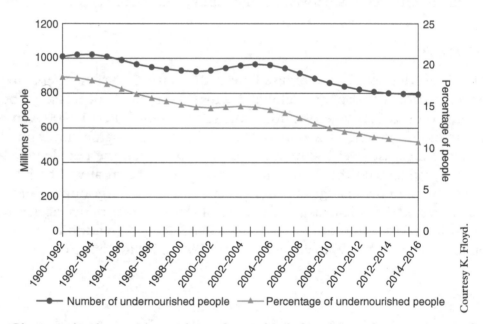

Figure 9.1. Changes in the total number of people (left axis) and percentage of population (right axis) who are undernourished. The World Food Summit (WFS) goal, created in 1996, was to reduce the *number* of undernourished people by half, to approximately 500 million, by 2015, while the Millennium Development Goal (MDG), created in 2001, was to reduce the *percentage* of undernourished people by half.

Data from FAO, http://faostat.fao.org/beta/en/#home.

Population growth and the expected continuation of a shift towards eating more meat as people become more wealthy (particularly in developing countries) leads to the prediction that the world agricultural production will need to increase 70% by 2050 (Bruinsma 2011). This projection does not include the potential increase in using crops to produce biofuels (refer to algae biofuels lab for more details), which could result in additional demands to increase overall agricultural production. Increasing food production requires increasing crop yields and/or expanding the area devoted to food production (Bruinsma 2011). Increases in yields can result from increased use of fertilizers and irrigation, both of which can have negative environmental impacts. Expanding the area devoted to food production is also challenging. Food production currently uses about a third of all of the land on Earth (UNEP 2014). About 70% of that is pastureland for raising livestock, with the remaining

30%, around 1500 million ha (Mha), used for growing crops (UNEP 2014). Much of the remaining land suitable for expanding agriculture is currently forested (mostly in Brazil and sub-Saharan Africa), so increasing cropland would come at the cost of additional deforestation and subsequent loss of biodiversity (Bruinsma 2011). And not all of the potential cropland is particularly well suited to agriculture, because of low soil fertility or terrain that is too variable (hills and valleys). In this unit we will focus on the issue of soil fertility, which is determined in large part by the soil properties, and the impacts that crop choice has on productivity.

The amount of organic matter in soils is a critical factor in determining soil fertility. Many agricultural practices tend to reduce the organic matter. Plowing soils disrupts soil structure and can accelerate the decomposition of organic matter. Removing the crops, along with the effects of plowing, exposes soils to wind and water erosion. Erosion removes the upper levels of the soil, which is where most nutrients and organic matter are found (Bot and Benites 2005), and soil fertility decreases. Even the basic act of harvesting crops for use in other locations removes nutrients and organic matter from the local system. Overgrazing in pasturelands and rangelands can also lead to increased rates of erosion when the livestock expose soil through feeding and disrupt soil structure by simply walking around.

Other human activities can decrease soil fertility as well. Poor irrigation practices can increase salt concentrations, particularly in arid and semi-arid regions. Salty soils create water stress in plants. Chemical pollutants can impact plant growth and can also accumulate in the edible parts of the plants, making them unsafe for consumption. Globally, about one third of land is moderately to highly degraded due to these human impacts.

In most agriculture, organic matter comes from leaving crop residues (stems, roots) in the fields. One example is conservation tillage, which aims to reduce the amount of plowing done relative to conventional agriculture. You can read more about sustainable agricultural practices in the Soil and Agriculture chapter in your textbook.

On a smaller scale, such as home gardens, adding compost can enhance soil fertility. Compost can be made from decomposed yard waste (leaves, grass cuttings). Kitchen scraps do not decompose well in typical compost piles, but make great food for worms. Worm composting uses kitchen scraps and produces rich organic matter for use in home gardens. Any type of composting also reduces the amount of organic waste that is disposed in landfills. In landfills, organic waste decomposes and produces methane, a potent greenhouse gas. Finally, using compost to grow more of your own food reduces the distance that the food has to travel, which reduces the carbon emissions associated with transportation.

In this unit, we will investigate redworm composting, how compost and inorganic fertilizers impact soil characteristics, and how those soil characteristics affect plant growth. Specifically, in the redworm compost lab we will calculate the amount of food scraps that redworms can compost weekly, and use that amount to estimate how less methane would be emitted from landfills if everyone composted. We will also observe which types of food scraps are more readily composted than others. In the soil characterization lab, we will compare physical and chemical properties of soil treatments (control, added inorganic fertilizers, added inorganic fertilizers and compost) using experimental plots housed on the Biology Green Roof. We will also compare the growth of crop plants among soil treatments. At the end of this unit you should have a greater appreciation for how to grow your own vegetables in a more sustainable way.

References

Bot, A., and Benites, J. 2005. *The Importance of Soil Organic Matter: Key to Drought-Resistant Soil and Sustained Food and Production. FAO Soils Bulletin 80*. Rome, Italy: Food and Agriculture Organization of the United Nations.

Bruinsma, J. 2011. "The Resources Outlook: By How Much Do Land, Water and Crop Yields Need to Increase by 2050?" In *Looking Ahead in World Food and Agriculture: Perspectives to 2050*, edited by Piero Conforti, 233–78. Rome, Italy: Food and Agriculture Organization of the United Nations.

Food and Agriculture Organization (FAO), International Fund for Agricultural Development and World Food Program. 2015. "The State of Food Insecurity in the World 2015." Meeting the 2015 international hunger targets: Taking stock of uneven progress. Rome, Italy: FAO.

United Nations (UN). 2015a. *Sustainable Development Goals*. www.un.org/sustainabledevelopment/. Accessed October 27, 2015.

United Nations, Department of Economic and Social Affairs, Population Division. 2015b. *World Population Prospects: The 2015 Revision, Key Findings and Advance Tables. Working Paper No. ESA/P/WP.241*. http://esa.un.org/unpd/wpp/Publications/. Accessed July 11, 2016.

United Nations Environmental Program (UNEP). 2014. *Assessing Global Land Use: Balancing Consumption with Sustainable Supply. A Report of the Working Group on Land and Soils of the International Resource Panel*. Lead authors: Bringezu S., Schütz H., Pengue W., O'Brien M., Garcia F., Sims R., Howarth R., Kauppi L., Swilling M., and Herrick J. www.unep.org/resourcepanel/Publications/AreasofAssessment/AssessingGlobalLandUseBalancing Consumptionw/tabid/132063/Default.aspx. Accessed July 11, 2016.

Resources

An overview of five strategies for how to feed 9 billion people sustainably. Short, easy to read, with some very good figures. www.nationalgeographic.com/foodfeatures/feeding-9-billion/

World Resources Institute article from 2013. The Global Food Challenge Explained in 18 Graphics. www.wri.org/blog/2013/12/global-food-challenge-explained-18-graphics

More in-depth reports from the World Resources Institute as part of the Creating a Sustainable Food Future project. www.wri.org/our-work/project/world-resources-report

wawritto/Shutterstock.com

Lab 9

Worm Composting

Learning Objectives

1. Understand the importance of organic matter for soil fertility
2. Describe how redworms help to compost organic matter
3. Explain how the amount and type of waste that goes into a landfill affects methane production and how composting can reduce it

Importance

Improving soil fertility is an important part of sustainably providing food for the growing human population. Soil organic matter provides nutrients to crop plants, improves water holding capacity, and acts to hold soils in place, reducing erosion. For commercial agriculture, techniques like no-till planting are cost-effective ways to increase organic matter but can have hidden costs such as the increased use of herbicides and pesticides. Backyard gardeners can use redworms to compost kitchen wastes, providing rich organic matter for gardens while reducing both the amount of organic matter being disposed in landfills, and the use of synthetic fertilizers. Decomposition of organic matter in landfills produces methane, a greenhouse gas more potent than carbon dioxide. Worm composting can help reduce carbon emissions associated with food transport if people to grow their own food using the composted wastes.

Guiding Questions

How much food do you throw away each week and how much could be used in redworm composting?

What are some ways that you can reduce food waste?

What are the advantages to growing your own food?

How does food production, transport, and disposal contribute to greenhouse gas emissions?

What other organisms are typically found in closed system compost bins?

Where and when are you going to set up your own redworm bin?

Vocabulary

1. *Red wriggler worms, redworms (Eisenia fetida)*—A type of earthworm that is well suited for home composting; they feed on decaying organic matter.
2. *Compost*—Mixture of decomposed organic material created under controlled conditions (i.e., not the product of decomposition in nature) that can be used to supplement soil quality.
3. *Municipal solid waste (MSW)*—Commonly known as trash or garbage, consists of items discarded, recycled, or composted by people. Does not include industrial, construction, or hazardous waste.
4. *Landfill*—Site were waste is isolated from the environment. Modern landfills have liners to prevent leaching of materials into the groundwater, and are generally covered with soil or something similar at the end of each day.
5. *Oxic*—Oxygen is present.
6. *Anoxic*—No oxygen is present.
7. *Food loss*—Food that is spilled or spoiled *before* it reaches the retail stage.
8. *Food waste*—Food that is discarded by retailers or consumers.

Introduction

Organic matter and soil fertility

As discussed in the introduction to the Food Production Unit, many current agricultural practices reduce soil fertility, largely by removing soil organic matter. Efforts to sustainably improve soil fertility focus on restoring and protecting organic matter. Several methods, such as contour farming and conservation tillage, are described in the Soil and Agriculture chapter of your textbook. On a smaller scale, such as backyard or community gardens, adding composted organic materials can increase the soil organic matter. Composting is essentially accelerated plant decomposition, and can be done with a mixture of leaves and grass clippings placed in a pile, kept moist, and turned occasionally to increase oxygen levels inside the pile. Kitchen scraps, like banana peels and apple cores, do not decompose as well in traditional compost piles, but are readily decomposed in red wriggler worm (*Eisenia fetida*, called redworms hereafter) compost bins as described as follows. With either composting method, the compost that is produced can be added to garden soils, improving the organic matter, soil fertility, and the ability of the soil to hold water.

Methane emissions from landfills and food waste

Americans generate about 250 million tons of municipal solid waste (MSW) per year, which consists of items we throw away, recycle, or compost, but does not include waste from factories, construction sites, or hazardous waste (Fig. 9.1A; EPA 2014). This works out to about 2 kg (4.5 pounds) of waste per person per day (EPA 2014). Of all the MSW produced each year, about 26% is recycled and another 8% is composted, diverting about 34% of MSW, around 85 million tons, from landfills (EPA 2014). Not all materials are recycled at similar rates. For example, about 67% of all paper products are recycled and 57% of yard waste is composted, but only 5% of food waste is composted (EPA 2014). In fact, once diversions from landfills by recycling and composting are considered, food waste makes up the largest single source of MSW in landfills (Fig. 9.1B; EPA 2014).

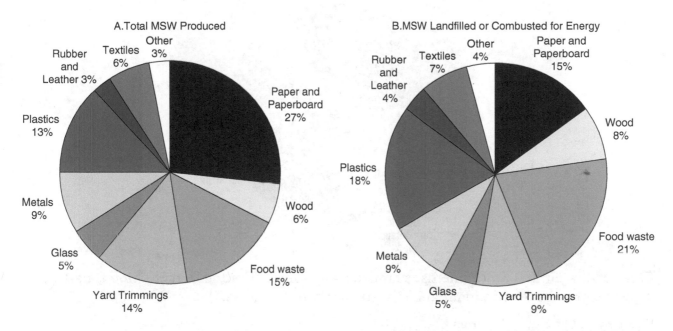

Figure 9.1. (A) Percentages of the major categories of MSW generated in the United States. About 250 million tons of MSW is produced each year. (B) Percentages of MSW that is sent to landfills or burned for energy recovery. The difference between the percentage produced and the percentage landfilled or combusted represents the amount recycled or composted. Overall about 34% of all MSW is recycled/composted, 12% is burned for energy, and 54% is sent to landfills.

Figures by K. Floyd with data from EPA (2014).

Some types of waste are biodegradable (also called organic waste), such as paper products, wood, food scraps, and yard waste. Together, these four categories represent 62% of all MSW produced (Fig. 9.1A; EPA 2014). When biodegradable wastes enter landfills, they are decomposed by bacteria. These bacteria produce "landfill gas." Landfills are often anoxic, particularly in the deeper levels. In the presence of oxygen, organic wastes decompose into carbon dioxide and water. However, in the anoxic parts of the landfill, organic wastes decompose into methane and water (see the Waste Management chapter of your textbook for more information on how landfills function). Overall, landfill gas is usually about 50% methane and 50% carbon dioxide, with small amounts of sulfur and nitrogen compounds (Matthews and Themelis 2007). Carbon dioxide and methane are both greenhouse gases, and concentrations of both have increased since the pre-industrial era (EPA 2006). Methane is more efficient than carbon dioxide at trapping infrared radiation (heat), and has a global warming potential of about 84 times that of carbon dioxide over a 20-year time frame. Thus, reducing methane emissions is an important component of slowing climate change. In the United States, methane emissions accounted for about 9% of all greenhouse gas emissions, and 17% of the methane emissions are attributed to landfills (Fig. 9.2; EPA 2015). Although it is possible to capture the methane from landfills and burn it as an energy source, we focus on reducing the amount of biodegradable materials, specifically food scraps, entering landfills as a way to reduce methane emissions to the atmosphere.

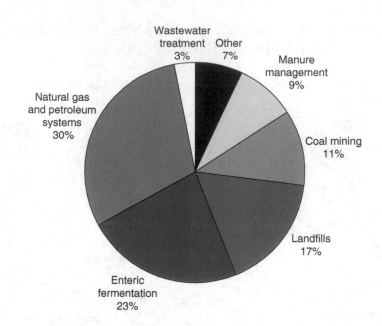

Figure 9.2. United States methane emissions by source. Enteric fermentation is part of the digestive process for ruminant animals, such as cattle or goats.

Figure by K. Floyd with data from EPA (2015).

Food waste and redworm composting

Reducing the amount of food that is sent to landfills begins with wasting less food. Globally, almost one third of all food produced is lost or wasted (Lipinski et al. 2013). In the United States about 40% of all food is wasted (Gunders 2012). This averages to 9 kg (20 pounds) per person each month (Gunders 2012)! Food can be lost or wasted at all stages of the supply chain, starting with losses in fields due to weather conditions to grocery stores discarding fruit considered too unattractive for selling to consumers throwing away expired food. Reducing these losses can actually save farmers and consumers money, along with reducing methane emissions from landfills.

Some foods have inherent waste, such as banana peels, apple cores, or coffee grounds. Using these materials, along with other suitable types of food scraps (Table 9.1), to feed to redworms instead of sending them to landfills is a way to reduce methane emissions while producing organic material called worm castings or worm compost. Worm castings can be used to improve the fertility of backyard lawns and gardens. Growing your own fruits and vegetables reduces environmental costs associated with processing and shipping food, often over great distances. In addition, compost bins are generally oxic, allowing aerobic decomposition and producing less methane than if the waste was sent to a landfill.

Redworm composting is easy and inexpensive. Redworm compost bins recreate soil food webs, with food waste providing the organic matter that makes up the first trophic level. The redworms eat the bacteria, fungi, and other small organisms that grow on and decompose food waste. The redworms also mix the contents of the bin as they move. Redworm bins should be given fruits and vegetables, but not meat or grains (Table 9.1). Although meat and grains do decompose, they tend to create unpleasant smells and can attract pests.

For this laboratory, each week a student will save their fruit and vegetable waste to bring to class for composting. Towards the end of the semester we will calculate how much food the redworms consumed, determine what types of foods seemed to decompose the fastest, and estimate how much reduction in methane emissions from landfills might occur if everyone composted their fruit and vegetable waste. We are also going to investigate biodiversity in the bins and how the addition of organic matter, including worm castings from our redworm bins, affects soil properties and the growth of plants grown on the roof of the Biology building (called the Green Roof, described in the Crop Production Lab portion of this unit).

Table 9.1: Types of foods that can and cannot be fed to redworms.

Good ☺	Bad ☹
• **Vegetable and fruit waste** • Everything except those listed in the bad column • Variety of items is better • These items should make up the bulk of the food waste • Rotten and moldy is great! • Chop items into smaller pieces • Add the following **in moderation** (no more than 25% of the food) • Egg shells (rinsed, dried, then crushed to as fine as possible—1 or 2 per week) • Coffee grounds (with the filter is OK) • Tea bags (remove staple, if present)	• Citrus, onions/garlic, large amounts of broccoli, cabbage, other veggies with similar smells • Dairy (cheese, milk, yogurt, etc.) • Meat • Grains/bread • Pet waste • Oils/grease • Nonbiodegradable materials (plastics, aluminum foil, Styrofoam, stickers on produce, etc.)

Activities

The lab compost bins have been in use for several years, so you do not need to set up the bins. In case you would like to set up your own bin, basic instructions are given in the Appendix at the end of this lab. In addition, there is a wealth of information about redworm composting available online. Two of the sites that we recommend are www.redwormcomposting.com/ and http://compost.css.cornell.edu/worms/basics.html.

Ongoing activities throughout the semester
  Feeding the redworms:

1. Every week, one person will fill a 1 L container with the *correct types of food waste* to bring to lab. Food will be frozen for a week to kill any fly or gnat larvae that might be present. Although these organisms do not cause problems in the composting, they can be a nuisance.
2. Take the food from the previous week out of the freezer to defrost at the beginning of lab.
3. Weigh the food. Record the total weight (in grams), the types of food, and its condition (chopped or whole, fresh or moldy, etc.) on the feeding log chart.

4. Move the bedding away from the next corner to receive food waste, add the waste, and then cover with bedding (see the following Figures). If the level of castings and bedding in the bin is too low to cover the food scraps, add more bedding.
5. Starting the fifth week, we will be adding food waste to a corner that had already received waste. After you move the bedding away from that corner, record what was happened to that food on the feeding log chart. Is all of it gone, or only some items? Which items are left?
6. Check the moisture level, and mist the bin if it is too dry. If it is too wet, add more dry bedding to the surface.
7. Repeat every week until the end of the semester.

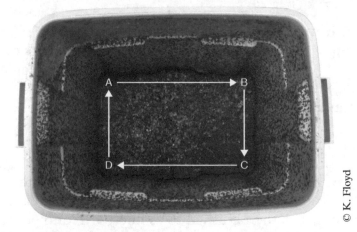

Each week add food to a different corner.

Move enough bedding out of the way to leave enough space for the food.

© Kevin Floyd

Add the food, and then cover completely with the bedding. Add more bedding if there is not enough to cover the food.

During the Redworm Composting lab
  Activity 1: Compost biodiversity
Composting involves a wide range of organisms, including bacteria, fungi, and invertebrates. Although our bins are indoors, and the only organisms added have been worms, there are still several species present. These species have introduced in the food scraps, the bedding, or through the air.

Place one spoonful of compost in a Petri dish, and examine it using a dissecting microscope (See Appendix 4: How to Use a Microscope). Record the number and types of organisms below. Use the following website to help in identifications: http://compost.css.cornell.edu/invertebrates.html. Also observe a redworm under the microscope, paying particular attention to their movement. Look for redworm egg cocoons (tan to brown, round to oval, and about the size of a match head). Each cocoon holds 1–5 eggs, which will hatch after 2–3 weeks.

Organism (e.g., springtail)	Description (e.g., small, brown)	Estimated abundance (#)

Activity 2: Redworm bin observations and methane calculations

Answer the questions in the lab assignment about patterns in food decomposition within the compost bin, total amount of food composted, and the potential reduction in landfill methane emissions possible if everyone used redworm composting bins.

Assessments

Complete the lab assignment available on Blackboard.

References

Environmental Protection Agency (EPA). 2006. *Solid Waste Management and Greenhouse Gases: A Life-Cycle Assessment of Emissions and Sinks.* http://nepis.epa.gov/Exe/ZyPURL.cgi?Dockey=60000AVO.txt. Accessed July 15, 2016.

Environmental Protection Agency (EPA). 2014. *Municipal Solid Waste Generation, Recycling, and Disposal in the United States: Tables and Figures for 2012.* https://www.epa.gov/sites/production/files/2015-09/documents/2012_msw_fs.pdf. Accessed July 15, 2016.

Environmental Protection Agency (EPA). 2015. *Overview of Greenhouse Gases: Methane Emissions.* http://epa.gov/climatechange/ghgemissions/gases/ch4.html. Accessed July 15, 2016.

Gunders, D. 2012. *Wasted: How America is Losing up to 40 Percent of its Food from Farm to Fork to Landfill.* New York: Natural Resources Defense Council. www.nrdc.org/food/wasted-food.asp. Accessed July 15, 2016.

Matthews, E. and Themelis, N. J. 2007. "Potential for reducing global methane emissions from landfills, 2000-2030." Sardina 2007, Eleventh International Waste Management and Landfill Symposium. Cagliari, Italy. http://www.seas.columbia.edu/earth/wtert/sofos/Matthews_Themelis_Sardinia2007.pdf. Accessed July 15, 2016.

Lipinski, B. et al. 2013. Reducing food loss and waste. Working Paper, Installment 2 of Creating a Sustainable Food Future. Washington, DC: World Resources Institute. http://www.worldresources-report.org. Accessed July 15, 2016.

Resources

The EPA's webpage on sustainable food management examines strategies for reducing food waste and for keeping waste out of landfills. It also has a section on what individuals can do. www2.epa.gov/sustainable-management-food, accessed May 8, 2016.

The Natural Resources Defense Council also has a lot of information about food waste. The author of the study referenced in the introduction published the book "Waste-Free Kitchen Handbook." www.nrdc.org/food/wasted-food.asp, accessed May 8, 2016.

The Natural Resources Conservation Service (NRCS, part of the US Department of Agriculture) has several webpages on soil biology, including soil food webs (www.nrcs.usda.gov/wps/portal/nrcs/detailfull/soils/health/biology/?cid=nrcs142p2_053868) and redworms (www.nrcs.usda.gov/wps/portal/nrcs/detailfull/soils/health/biology/?cid=nrcs142p2_053863), accessed May 8, 2016.

There is a lot of information available online about redworm composting (also called vermicomposting). Searching the internet for either term will yield lots of interesting results.

Appendix

Setting up a compost bin:

1. Get a plastic container, roughly 60 × 30 × 25 cm (or larger). Air flow is important for the worms and the other decomposers, so several holes must be drilled into the lid and the sides. In addition, most food waste tends to be fairly moist, so drainage holes are needed in the bottom of the container. If the holes are not already present, make them using a sharp screwdriver or drill (about 0.3 cm [1/8″] diameter, a couple inches apart).

2. Label the corners A–D. This helps when tracking where food was added and where to add food next.

3. Line the bottom of the container with a sheet of newspaper. This will help keep the worms in place as they are adjusting to their new home.

4. Fill the container about a third full with bedding (e.g., shredded newspaper or cardboard, coconut fiber called coir, etc.), to a depth of about 15 cm. White office paper should not be used because the bleaching process used during its manufacture can be harmful to worms.

5. Scatter a handful of sand (or garden soil that has not been treated with pesticides) in the bin. Sand is swallowed by the worms and used in their gizzards to grind food.

6. Wet the bedding with water until it has the moisture content of a wrung-out sponge.

7. Pull the bedding away from a corner (A) and add the first set of previously-frozen food waste. Cover the waste with bedding.

8. Put the lid back on the container and wait a week. This allows the waste to begin to decompose before adding the worms.

9. After waiting the week, add the worms to the bin. A typical bin of the size recommended can be started with 1000–2000 red wiggler worms (*Eisenia foetida*), a type of earthworm. You can start with a smaller population, but would need to reduce the amount of food waste added until the redworm population increases. You can purchase red wiggler worms at pet stores, but for larger quantities it is more economical to order them online. The other common species of commercially available earthworms called night crawlers (*Lumbricus terrestris*) do not work well in compost bins because they prefer deep burrows.

© K. Floyd

Taos red beans growing on the Biology
Green Roof at UTEP.

Crop Production

Learning Objectives

1. Understand potential benefits of eating locally produced and/or heirloom or native-adapted food
2. Relate growth and reproduction of plants to soil type and fertilizer type
3. Compare growth and reproduction of native-adapted and commercial varieties of similar vegetables
4. Construct and analyze graphs

Importance

Modern industrial agriculture has increased food production over the past half century, but at a large environmental cost. Alternative food production systems include producing food locally, growing heirloom or native-adapted crop varieties, and improving the soil quality with organic matter. These alternative systems have the potential to produce high quality food while reducing some of the environmental costs of industrial agriculture.

Guiding Questions

Where is your food produced?

How does soil quality and nutrients affect plant growth?

What vegetable varieties are best suited to our local environment?

What steps would you take if you wanted to begin growing your own vegetables at home?

Vocabulary

1. *Green Revolution*—The intensification of agriculture that started in the mid-twentieth century, uses large quantities of irrigation water, synthetic fertilizers and pesticides, and crops bred specifically for high yield.
2. *Fertilizer*—Something that enhances plant growth by providing nutrients.
3. *Pesticide*—A chemical that kills unwanted pests, such as insects (insecticide), weeds (herbicide), or fungi (fungicide).
4. *Monoculture*—The planting of a single species or variety of crop over a large area.

5. *Organic agriculture*—A system of agriculture that uses no synthetic fertilizers or pesticides, but instead relies on biological agents and soil improvement.
6. *Heirloom crop varieties*—Varieties that are naturally pollinated and predate breeding programs used by modern agriculture. More common in backyard gardens or small-scale agriculture than on large farms.
7. *Farmer's markets*—Areas where multiple farmers can sell their produce, meats, or prepared foods directly to customers.
8. *Community-supported agriculture (CSA)*—A system where farmers sell shares of their production to consumers, generally before the planting season begins. The consumers receive their shares of produce or meat throughout the season.
9. *Cultivar*—A specific variety of one species that has been selected for desirable characteristics, such as high yield, better flavor, or disease resistance.

Introduction

Modern industrial agriculture, including the Green Revolution that began in the 1940s, has done a remarkable job of increasing food production, helping to reduce both the absolute number and relative percentage of undernourished people since 1990 (discussed in the Introduction to this unit). Industrial agriculture requires large inputs of irrigation water, synthetic fertilizers, chemical pesticides, and fossil fuel-driven machinery like tractors. It also tends toward monocultures and specific varieties of genetically similar crops. These changes have increased the average crop yield per hectare, allowing a large increase in food production while converting a relatively small amount of additional land to agriculture. However, these increases have other environmental costs, including eutrophication of local water bodies and dead zones from increased use of chemical fertilizers (see the Water Quality lab for more information on this issue), greenhouse gas emissions from several sources: the use of fossil fuels, nitrogen fertilizers (which produce nitrous oxide, N_2O), and confined livestock operations (which produce methane, CH_4), loss of fertile topsoil, and loss of biodiversity from pesticide use and specially-bred crops. Alternatives such as organic agriculture use methods that try to maintain healthy soil, clean water, and a stable climate system. *We will investigate three specific practices that attempt to improve the sustainability of the food system: locally produced food, heirloom plant varieties, and use of compost to improve soil fertility.* Compost is discussed in more detail in the Worm Composting lab in this unit, while soil fertility is discussed in the Soil lab and the Introduction to this unit. The Soil and Agriculture chapter of your textbook also discusses these and other issues related to food production.

Locally produced food

The 2008 Farm Act defined food to be "local" if it was produced within 650 km (400 miles) or in the same state as where it is consumed (Martinez et al. 2010). Locally produced food can be purchased at some grocery stores, or directly by a consumer at a farmers' market, through CSA, or at farm stands. Here in El Paso we have the Ardovino's Desert Crossing Farmers' Market, which started in the early 2000s, and the Downtown Art and Farmers Market, which started in 2011. Even though the number of farmers' markets and CSAs in the United States has more than doubled over the past 20 years, and accounts for over

$1 billion in sales, these direct-to-consumer sales are still less than 1% of total agricultural sales (Martinez et al. 2010). Personal gardens, including backyard and community gardens, are another approach to obtaining locally grown foods. In 2009, 43 million US households, about a third of all households, intended to grow their own produce (Martinez et al. 2010). This food production was estimated to be worth over $2.5 billion (Martinez et al. 2010).

The main environmental benefit of locally grown food is the potential of lower carbon emissions due to transportation of food. In the global food system, items can be transported large distances from where they were grown to where they are consumed. Many stores provide the same varieties of fresh produce year-round, despite the fact that the most crops have a distinct growing season. For example, cherries and strawberries ripen in spring and summer in North America, but can be grown during the North American winter in Central and South America and shipped over long distances to stores in the United States. Buying local and in-season produce can reduce the overall carbon footprint associated with food. However, the greenhouse gas emissions associated with food are mostly from the production phase, with only about 20% of the total emissions coming from the transportation of the food to the consumer (Weber and Matthews 2008). Weber and Matthews (2008) found that switching from high-greenhouse gas associated foods like meat and dairy to chicken, fish, or vegetables once a week would offset the transportation-related emissions. Other benefits to buying locally produced food include supporting local economies, getting fresher food, encouraging healthier food choices, and increasing resilience in the food system by supporting a larger number of smaller producers (Martinez et al. 2010).

Heirloom and native crop varieties

The Food and Agriculture Organization (FAO) estimates that 90% of the food we consume globally comes from just 15 crop species and eight livestock species (FAO 1999). Rice, maize (corn), and wheat alone contribute about 60% of calories we consume (FAO 1999). Industrial agriculture promotes the production of a small number of genetically similar crops. These genetic strains tend to produce high yields and increase the efficiency of growing and harvesting the crops. However, they are also at greater risk to new pathogens and changing environmental conditions. Genetic diversity in a variety of cultivars or in the wild relatives of crop species can be useful in breeding programs that can introduce beneficial adaptations (e.g., disease resistance, drought tolerance) to the domesticated crop species (Esquinas-Alcazar 2005).

The genetic diversity of crop species can be conserved through the increased use of heirloom varieties or native crops. Heirloom plants are generally older cultivars that pre-date modern breeding practices (roughly pre-1950), and are open pollinated, meaning that their seeds can be used to grow the same cultivar in following seasons (something not possible with most modern crop varieties; Russ and Bradshaw 1999). In addition to maintaining genetic diversity, many heirloom varieties taste better than the commercial varieties! Native-adapted crops are similar to heirlooms, but are plants that have been cultivated within a region by indigenous cultures since prehistoric times (Nabhan 1995).

Native crops have been grown in a region long enough to have adaptations to the particular soil and climate, and can potentially be grown with fewer inputs like irrigation water and fertilizers.

Experimental design

We planted various crop species in planters on the roof of the Biology building at the beginning of the semester. We chose beans and squash or spinach, depending on the season, as our crops. Beans and squash, along with corn, are the three main crops of Native American groups in North America (Landon 2008). When planted together, these species are often called the "three sisters." The corn stalks provide support for the climbing beans, the beans provide nitrogen to the soil, and the squash provides soil cover that reduces water loss and prevents weed growth. Combined, corn, beans, and squash contain essential fatty acids and all essential amino acids. However, our planters are too shallow for the deep-rooted corn to grow, so we are not including that species in the experiment. We planted both native-adapted and commercial varieties of each species in individual plots (although these might change from semester to semester; Table 10.1). There are three soil treatments for the native-adapted plants: control ("0"), inorganic fertilizer ("NP"), and compost with inorganic fertilizer ("COM"). The commercial varieties are only grown in NP soil. The basic soil in the planters is a mixture of red lava rock and fine grain sand. Compost added was a combination of commercially available composted mulch, locally produced compost made in part from animal manure and bedding from the El Paso Zoo, and redworm castings from the worm bins in the ESCI lab. Each semester more redworm castings harvested from the worm bins are added to the compost treatment plots. Read the general Introduction and the Soil lab in this unit for more information about soil fertility.

Table 10.1: The species and varieties grown for this experiment.

Species	Native-adapted variety	Commercial variety
Beans	Taos red bean	Pinto beans
Squash	Tarahumara squash	Jack O'Lantern pumpkin
Spinach	Tarahumara espinaca	Correnta hybrid

Our research questions are (1) How does soil treatment affect the growth of native-adapted plants, and (2) Do native-adapted plants grow better than commercial varieties? Based on this introduction and any outside reading you have done, what is your hypothesis for each question?

Hypothesis 1:

_____

Hypothesis 2:

_____

You will attempt to answer these questions by measuring the plants and making graphs that show the size and reproductive output of the soil treatments. You will make some conclusions based on those results, and think about ways that the experiment could be improved in the future.

Activities

Plant measurements

1. Each group will measure the plants in 6–8 plots.
2. **Leave the plant as it is for the measurements** (i.e., do not pull it up when measuring the height or try to straighten it when measuring its circumference).
3. Measure the height (cm) from the surface of the soil to the tallest part of the plant.
4. Measure the area covered using one of the following methods:
    a. If the plant is relatively small, use a grid to estimate the area covered by the plant. The grid has wires spaced approximately 2 cm apart, so each square is around 4 cm^2 (Fig. 10.1). Chose legs that hold the grid as close to the plant as possible without touching any of the leaves. Looking down from directly above the plant, count the number of squares that you can see the plant through. Multiply the number of squares by 4 to get an estimate of the area covered. This is an overestimate because not every square will be completely filled by the plant, but it is a relatively fast method to approximate the area covered.
    b. If the plant is relatively large, measure the length and width of the plant in cm (not touching the plant) and assumes that the overall shape of the plant is an ellipse (similar to an oval). The formula for calculating the area of an ellipse is $A = \pi \times$ (length/2) × (width/2). You can use Excel to easily calculate the area for all of the plants after they are measured.
5. Measure the reproductive output by counting the number of flowers and the number of fruits (bean pods or pumpkins/squash).

What alternative methods do you think would improve the accuracy of the size estimates?

Each plant is numbered using plastic sticks. We planted six beans, six spinach seeds, and/or three squash seeds near the beginning of the semester. However, not all the seeds germinate, and some plants may die before we are able to complete this lab. For that reason, there are 2–6 bean or spinach plants per plot, and 1–3 squash per plot. When recording the data, make sure that you are entering the data for the correct plot and the correct plant. **Be careful not to step on any of the plants or damage them during the measurements.** Try to avoid walking on the soil as much as possible.

Data analysis

1. Combine the data from all of the groups into a single complete set. Use Excel to calculate the average and standard deviation of the (1) height, (2) area covered, (3) number of flowers, and (4) number of fruits for each soil treatment, keeping the native-adapted and commercial varieties separate.
2. Graph the average height (spinach) **or** average area covered (beans and squash) of the native-adapted variety only in each of the three soil treatments. Include the standard deviation as error bars.

© K. Floyd.

Figure 10.1. Example of how to use the grid to estimate the area covered by a plant. Each square is approximately 4 cm². Count the number of squares that have any part of the plant showing and multiply by 4. Although this is an overestimate of the area covered, it is a relatively fast method.

3. Graph the average height (spinach) **or** average area covered (beans and squash) of the native-adapted versus the commercial variety, using only the native-adapted plants from the NP soil treatment. Include the standard deviation as error bars.
4. Note that each lab section will not necessarily complete measurements on both species of plants. Your TA will tell you which plants your section will be measuring in class.

Assessments

Copy and paste the two graphs into a single Word document. Print out one copy per lab group and write the names of all the group members on the printout. Complete the lab assignment posted on Blackboard individually.

References

Esquinas-Alcazar, J. 2005. "Protecting Crop Genetic Diversity for Food Security: Political, Ethical and Technical Challenges." *Nature Reviews Genetics* 6: 946–53.

United Nations Food and Agricultural Organization (FAO). 1999. *Women: Users, Preservers and Managers of Biodiversity*. www.fao.org/docrep/x0171e/x0171e03.htm. Accessed July 15, 2016.

Landon, A. J. 2008. "The 'How' of the Three Sisters: The Origins of Agriculture in Mesoamerica and the Human Niche." *Nebraska Anthropologist* 20: 110–24. http://digitalcommons.unl.edu/nebanthro/40. Accessed July 15, 2016.

Martinez, S., Hand, M., Michelle Da Pra, Pollack, S., Ralston, K., Smith, T., Vogel, S., Clark, S., Lohr, L., Low, S., and Newman, C. 2010. *Local Food Systems: Concepts, Impacts, and Issues*. ERR 97, U.S. Department of Agriculture, Economic Research Service.

Nabhan, G. P. 1985. "Native Crop Diversity in Aridoamerica: Conservation of Regional Gene Pools." *Economic Botany* 39 (4): 387–99. www.jstor.org/stable/4254790.

Russ, K. and Bradshaw, D. 1999. *Heirloom vegetables. HGIC 1255, Clemson Cooperative Extension.* www.clemson.edu/extension/hgic/plants/vegetables/gardening/hgic1255.html. Accessed July 15, 2016.

Weber, C. L. and Scott Matthews, H. 2008. "Food-miles and the Relative Climate Impacts of Food Choices in the United States. *Environmental Science & Technology* 42 (10): 3508–13.

Resources

A story about the Ardovino's Farmers Market from 2014: https://fnsnews.nmsu.edu/border-farmers-market-a-year-round-hit/

The press release from a 2015 study that estimated that up to 90% of Americans could obtain all of their food from producers within 100 miles of their homes: www.ucmerced.edu/news/2015/most-americans-could-eat-locally-research-shows

Video about the world's first rooftop farm. Available through the UTEP library at http://0-fod.infobase.com.lib.utep.edu/PortalPlaylists.aspx?wID=104347&xtid=56744.

Native-adapted seeds for the Southwestern U.S. can be ordered from Native Seeds: http://www.nativeseeds.org/

www.shutterstock.com

Lab 11
Soils and Agriculture

Learning Objectives

1. Relate soil properties to food production
2. Describe how environmental factors influence the type of soil found in a particular location
3. Categorize soils by color, texture, and nutrient composition
4. Explain how soil texture influences movement of water through soil

Importance

Food production depends heavily on the quality of the soil. Soil provides nutrients, water, and support to plants. Unfortunately, many current agricultural practices, such as plowing, leaving fields without plant cover during part of the year, and flood irrigation, actually degrade the soil. It is important to understand, how soils are formed and how soil properties influence plant growth. Protection of soils is one key factor in being able to sustainably provide food to a growing human population.

Guiding Questions

How does food production impact the environment?

Why are certain regions of the world called "bread baskets" while others produce very little food?

Why is soil often considered to be a nonrenewable resource?

Vocabulary

1. *Arable land*—Land suitable for growing crops. Of the approximately 5000 million ha used for agriculture, about 1500 million ha (30%) are arable.
2. *Parent material*—The materials from which the mineral components of soil are formed. It can be the underlying rock or materials transported to a location by wind or water erosion.
3. *Weathering*—The breakdown of rocks and minerals into smaller particles. It can be mechanical, where the rock crumbles into smaller parts through actions like the freeze-thaw cycle, or chemical, where the materials that make up the rocks are changed. Examples of chemical weathering are the dissolution of limestone or the rusting of iron minerals.

4. *Erosion*—The transport of materials through either wind or water.
5. *Leaching*—The loss of water-soluble minerals from the soil into water. The minerals can be carried into deeper soil horizons or potentially into groundwater or evaporated on the surface.
6. *Topography*—The degree of slope of the land, and its overall three-dimensional shape.
7. *Porosity*—The amount of open space in between soil particles. The pores can be filled with air or water. If all the pores are filled with water, the soil is considered saturated.
8. *Field capacity*—The amount of soil moisture held in the soil after the excess water has drained away. It is a measure of the soil water-holding capacity. When a soil is at field capacity, the pores between soil particles contain both water and air.

Introduction

Soil and agriculture

Soil is a complex mixture of minerals, organic matter, air, water, and living organisms such as bacteria, fungi, arthropods, and plant roots. Soil formation and characteristics are discussed later. Soil fertility refers to the nutrients that plants can access and suitable soil structure to allow root growth to both reach nutrients and anchor the upper portions of the plant (the shoot system). Highly fertile soils contain both macronutrients such as nitrogen, phosphorus, and potassium, as well as micronutrients, such as copper, iron, and zinc. Organic matter (derived from on-site decomposition of animal or plant matter) can help provide food to soil organisms, increase soil fertility, and improve the water-holding capability of the soil (Bot and Benites 2005). Organic matter is usually only 2%–10% of soil, but is very important for soil fertility (Bot and Benites 2005). Many agricultural practices, such as plowing, leaving fields bare, and overgrazing, tend to reduce soil organic matter, and thus soil fertility.

There are agricultural methods that avoid negative impacts on soil fertility, mostly involving ways to increase the amount of organic matter in the soil (Bot and Benites 2005). Specific methods for commercial agriculture such as intercropping and conservation tillage are described in the Soil and Agriculture chapter of your textbook. In smaller fields or home gardens, adding compost can increase the organic matter. Compost is discussed in the Worm Composting lab. Another common approach to increase the nutrients available to plants is to add chemical fertilizers. Chemical fertilizers are readily available to plants, and can stimulate improved growth within days of application. However, they do not add organic matter, and excess application of fertilizers can contribute to nutrient pollution in local waterways. Nutrient pollution can trigger algal blooms that can create dead zones. See the labs on Water Quality and Algal Biofuels for more information.

Agricultural production also depends on the climate. Areas that are too dry, too hot, or too cold have low production of crops. The areas of the world that produce the most food have the best combination of soil and climate, and are often referred to as "breadbaskets." This interaction of plants and climate is also important in the formation of soil.

Soil formation and classification

There are five main factors that influence soil formation: parent material, organisms, climate, topography, and time. These factors interact and do not operate in isolation. *Parent materials* are the mineral components that become soil. They can be weathered rocks (Fig. 11.1) or material transported to a location via wind or water. *Organisms* contribute organic matter as discussed earlier. Animals and plant roots can also create openings in soils that help with air and water flow. Some organisms, like lichens, can actually increase the rate of rock erosion into smaller particles by producing acidic secretions. The *climate* can influence rates of weathering and leaching, as well as wind and water erosion. Climate also impacts the types of organisms found in a region. The *topography* can influence the soil moisture and temperature, with areas lower on a slope generally having more moisture as the precipitation moves downslope. All else being equal, soil is more likely to erode on steep slopes than on flat ground. Slopes that face away from the sun (north slopes in the Northern Hemisphere) also tend to be wetter and cooler. Finally, all of these processes take *time* to occur. Soil on recently exposed parent material, such as might be exposed during a rock slide or flood, will look very different in 100 or 1000 years (Fig. 11.2). Thus soil is continuously created, but many modern agricultural practices lead to erosion rates that far exceed the rate of soil creation.

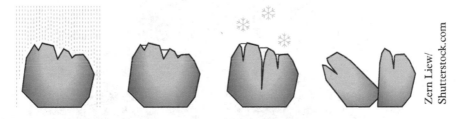

Zern Liew/
Shutterstock.com

Figure 11.1. Weathering from rain and freezing breaks larger particles into smaller particles. The size of the particles determines whether they will be classified as sand, silt, or clay.

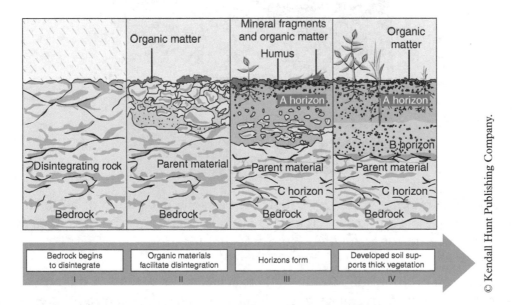

© Kendall Hunt Publishing Company.

Figure 11.2. Formation of soils. The initial colonization of the parent material by plants or lichens introduces some organic matter, which in turn facilitates the growth of other types of plants. Over time, this can lead to formation of different soil horizons. This process is an example of primary succession.

As soil forms, it develops layers called horizons (Fig. 11.3). The top layer is called the O horizon and consists of freshly decomposed plant residues (e.g., leaves). The A horizon is a mixture of organic matter from the surface and minerals from the weathered parent material below. This horizon is commonly called topsoil and generally has the highest fertility and the majority of plant roots and animals. The B horizon is the partially weathered parent material with very little organic matter. It is often called subsoil, and in some areas, the minerals that leach from the A horizon accumulate in the B horizon. The C horizon is the parent material. The R horizon (also called D horizon) is the bedrock (when present), and is composed of large sections of rocks.

SOIL LAYERS

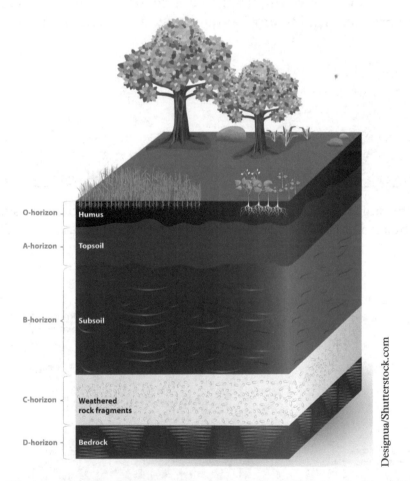

Figure 11.3. The major soil horizons that form as soils develop. Not all locations will have all horizons (e.g., desert soils often lack an O horizon), and the thickness of each horizon will depend on the specific soil formation factors at that location.

The interaction of these soil formation factors leads to a variety of soil types. Just as biologists classify species based on similarities, soil scientists classify soil into different orders. There are several important characteristics used to classify soils. If you are examining the soil at a particular site (in situ), you can assess the *structure* of the peds (naturally occurring aggregates of individual soil particles, similar to clods, which are human-formed soil fragments). Different structures are granular, blocky, prismatic, columnar, or platy (see GLOBE diagram posted on Blackboard for details). Some soils do not have structure, and are either single grained (common in sandy soils) or massive (usually high clay concentrations). The *color* of the soil can indicate levels of organic matter (dark brown or black), iron (red or orange), or calcium and silica (gray or white). These are useful for classification and comparisons, but additional testing would be required to measure the actual amounts of these materials in the soils. The soil *consistence* measures how difficult it is to break apart individual peds. Typically, soils with higher clay content are firm, while loams are more friable. The soil *texture* is determined by the different amounts of the three types of mineral particles: sand, silt, and clay. Sand particles are 0.005–2.0 mm in diameter, and sandy soils have a gritty feel. Silt particles are 0.0002–0.005 mm and silty soils feel smooth, like flour. Clay particles are <0.0002 mm in diameter, and clay-rich soils feel smooth and slick. Once the relative amounts of each type of particle in a soil sample are determined, the soil texture triangle (Fig. 11.4) can be used to determine the texture class. We are going to use a different method to determine the texture class, based on the feel of the soil. We discuss the importance of soil texture in more detail later.

The final important characteristic is soil *chemistry*. In particular, the presence of free carbonates, the pH, and the nutrients are often measured. Free *carbonates* are present in soils derived from carbonate-rich parent material, like limestone. Carbonates act as a buffer against changes in pH, and soils with carbonates are often basic (pH > 7). Farmers will often add carbonate-containing compounds to help neutralize soil acidity, especially in locations like the eastern United States that have acid deposition. The *pH* can impact the availability of nutrients to the plants, and thus influences what types of plants can grow in soils. Human actions

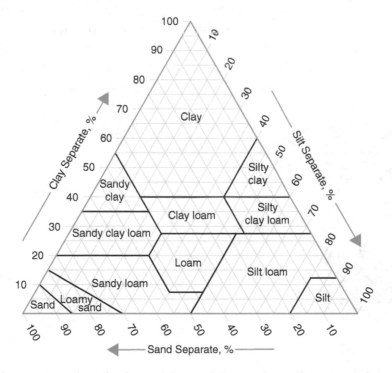

Figure 11.4. Soil texture triangle from Natural Resources Conservation Service (NRCS). Once the relative percentages of each size class is known, the triangle can be used to determine the soil texture. For example, a soil that is 40% sand, 40% silt, and 20% clay is a loam. The triangle can also be used in the opposite direction, by determining the relative proportions of the particles once the texture class is known.

Courtesy USDA www.nrcs.usda.gov/wps/portal/nrcs/detail/soils/edu/?cid=nrcs142p2_054311

can modify the soil chemistry, especially nutrients if the soil is used for agriculture. The three *nutrients* most commonly monitored are nitrogen, phosphorus, and potassium, abbreviated NPK after their chemical symbols. Together, these characteristics can inform us about how the soil formed, what type of ecosystem it can support, and what type of crops can be grown.

There are 12 soil orders and many more subgroups (USDA, Soil Formation and Classification). Although the details of soil orders are beyond the scope of this class, we will examine a couple to highlight how the formation factors lead to distinct soil types. Desert soils are often light colored due to a lack of organic matter (Fig. 11.5). The low amount of precipitation reduces plant productivity and also reduces the amount of chemical leaching from the soils. Accumulation of calcium carbonate in the B horizon can create the hard layer of *caliche* commonly encountered around El Paso. Grassland soils have a dark A horizon from the accumulation of plant organic matter (Fig. 11.6). These form in regions with moderate precipitation to support high productivity of grasses, but with sufficient fires or grazing by large animals to prevent trees or large shrubs from colonizing. Many of the grasslands have been converted to agriculture because of the fertile soil (Fig. 11.7).

Soil texture depends on the relative amounts of sands, silts, and clays. Soil texture affects the way that water infiltrates into the soil, how long water is retained in the soil, and how plant roots can grow through the soil. This is caused by the inverse relationship between soil texture and pore spaces (Fig. 11.8). Although soil porosity depends on both soil texture and soil structure (how the individual particles are aggregated into peds), here we focus on texture. Soils that have more sand-sized particles will have larger pore spaces, which allows air, water, and organisms to move through the soil easily. Soils that have more clay-sized particles have small pore spaces, which means air, water, and organisms have a harder time moving though the soil. Increased water infiltration in coarse soils can also increase leaching of nutrients from the active layer to deeper layers where plant roots have a harder time accessing them. The smaller clay particles have very high surface areas, which increases their

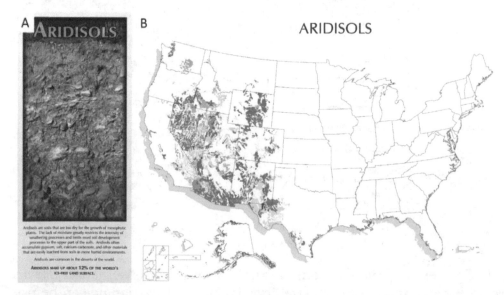

Figure 11.5. (A) Typical profile of a desert soil ("aridisol"). Note the general lack of organic matter in the A horizon and the accumulation of white salts in the B horizon. (B) Map of the distribution of aridisols in the United States. The shades represent different suborders of aridisols. Both images from USDA-NRCS.

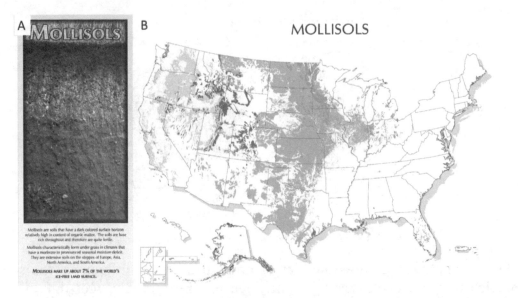

Figure 11.6. (A) Typical soil profile of a grassland soil ("mollisol"). Note the dark organic-rich A horizon. (B) The distribution of mollisols in the United States. The shades represent different suborders of mollisols. Both images from USDA-NRCS.

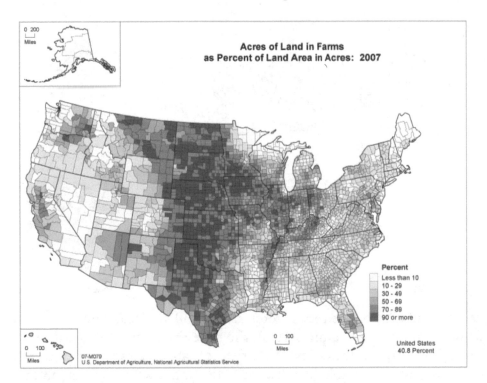

Figure 11.7. Distribution of farmland in the United States. Note the overlap between farmland and location of mollisol soils (Figure 11.6B). Image from USDA-NRCS. (http://www.agcensus.usda.gov/Publications/2007/Online_Highlights/Ag_Atlas_Maps/Farms/Land_in_Farms_and_Land_Use/07-M079.php)

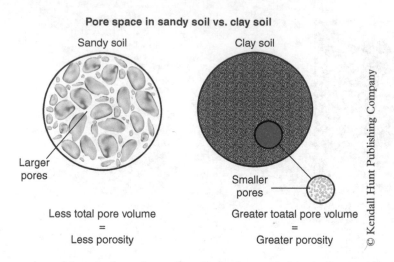

Figure 11.8. Schematic of how the size of mineral particles influences the soil porosity. Larger particles do not pack together tightly, leaving more pore spaces that air, water, and organisms to move through.

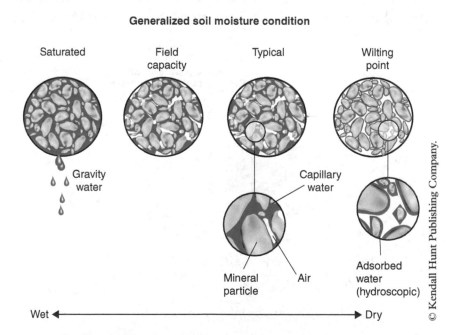

Figure 11.9. An illustration of the differences in soil moisture between saturation, field capacity, and wilting point. At saturation, all the pore spaces are filled with water, and water will drain through the soil to lower levels. At field capacity, some of the pores are filled with air, but there is still water available for plant growth. At the wilting point, all of the remaining water is tightly adsorbed to the soil particles, and is not accessible by plants.

ability to hold water and nutrients via adhesion (increasing their field capacity, the amount of water retained in the soil—Fig. 11.9). In general, the higher drainage rates of sandy soils makes water and nutrients less available for plants, but does allow for air flow and relatively easy routes for roots to grow. Soils with high clay content drain more slowly, making water and nutrients available for a longer time. But these soils can become waterlogged, and it is more difficult for plant roots to grow through the smaller pore spaces. A mixture of 40%

sand, 40% silt, and 20% clay particles is the texture class called loam (Fig. 11.4). Loams are ideal for most plants because they balance water and nutrient retention with soil drainage. For example, the mollisols where much of the agriculture currently occurs have a silty loam soil in the A horizon (Figs. 11.6 and 11.7; USDA 1999).

Activities

Today, we will characterize soils focusing on their physical and chemical properties. We will examine five samples, three from the Biology Green Roof (control, inorganic fertilizer, and inorganic fertilizer plus compost) and two from the region (desert mesa and river flood-plain). Your group will need to assess the *color, consistence, texture,* and *free carbonates* for *each soil sample.* Because we do not have the soils in situ, we will not be assessing the overall soil structure. Your group will measure the *pH, nitrate, phosphate, potassium, soil permeability and field capacity,* and *soil texture with the settling method* for a *single sample,* and then share those results with the rest of the class.

Following are the instructions for the activities. Perform the activities in the following order to maximize the use of the lab time.

1. Sieve the dried sample with a 2 mm sieve. Use the soil that passes through the sieve, the <2 mm size fraction, for the tests.
2. Start the water extraction of H^+ ions and nutrients by adding 50 g of soil and 100 mL of deionized (DI) water to a 250-mL centrifuge bottle. Cap and shake for 30 seconds. Let sit for about 5 minute, then shake for 30 seconds. Repeat the shake and wait cycle three additional times.
3. During the waiting periods between stirring the soil/water solution, measure the color, consistence, texture, and free carbonates of all the soil samples.
4. After the 20–30 minutes of the extraction, measure the pH using the pH meter.
5. Add or remove supernatant until all of the samples weigh 225 g (with the lid).
6. Centrifuge the samples in the Bioscience Core Facility. It will take around 20 minutes to complete. Filter the supernatant, and then transfer into a clean beaker.
7. Use the supernatant of the centrifuged sample to measure nitrate, phosphate, and potassium (NPK) using the YSI water chemistry kits.
8. During the waiting periods while conducting the water chemistry measurements, measure the soil permeability.
9. Again, during the water chemistry waiting periods, set up the soil texture by differential settling demonstration. You will mark the sand fraction today, and then look at the silt and clay fractions next week.
10. Share the results with the rest of the class.

Record your data on the lab assignment as you go through the following activities.
Instructions modified from Soil Characterization Protocol from globe.gov.
For All Soil Samples
Color: Moisten the sample slightly if it is dry. Find the color in the Munsell soil color guide that most closely matches the color of the sample. Record the color description in words, and not the hue/value/chroma code (e.g., do not write down "10R 4/4," write "reddish brown").

Consistence: Take a ped from the soil sample. Holding the ped between your thumb and forefinger, gently squeeze until it falls apart. Record one of the following categories on the datasheet:

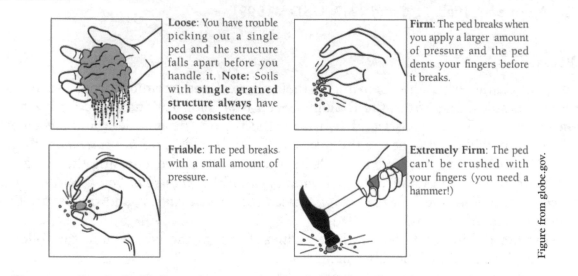

Loose: You have trouble picking out a single ped and the structure falls apart before you handle it. **Note:** Soils with **single grained structure always** have **loose consistence.**

Firm: The ped breaks when you apply a larger amount of pressure and the ped dents your fingers before it breaks.

Friable: The ped breaks with a small amount of pressure.

Extremely Firm: The ped can't be crushed with your fingers (you need a hammer!)

Figure from globe.gov.

Texture (by feel): Follow the steps in the following flowchart, and record the results on the data sheet:

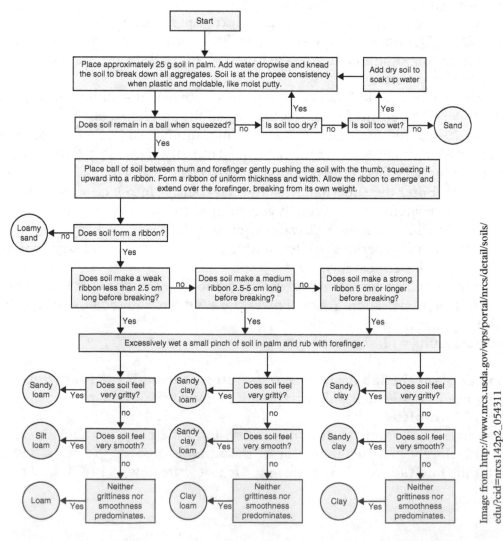

Image from http://www.nrcs.usda.gov/wps/portal/nrcs/detail/soils/edu/?cid=nrcs142p2_054311

Free carbonates: Use a pipette to add several drops of vinegar to a portion of the soil sample. Vinegar is a weak acid, which will react with carbonates in the soil to produce carbon dioxide bubbles. Record the results as none (no reaction, meaning no free carbonates present), slight (some bubbling), or strong (many and/or large bubbles, indicating many carbonates present).

For a single soil sample:
Chemical properties:

1. Extract soil nutrients and hydrogen ions (to measure pH) into a solution. In a 250-mL centrifuge bottle, mix 50 g of dried, sieved soil, and 100 mL of distilled water. Shake for 30 seconds, wait for at least 4 minutes. Repeat the shake and wait cycles four additional times, then wait about 5 minutes for the mixture to settle (total of about 20–30 minute).
2. Measure the pH using a pH meter. Be sure to rinse the bulb with DI water before and after the measurement. Place the bulb into the beaker of tap water in between measurements.
3. Add or remove supernatant until the bottle (with lid) weighs 225 g. The TA will centrifuge the samples in the Biosciences Core Facility to help pellet the suspended solids.
4. Once centrifuged, filter the supernatant to remove any floating organic particles and suspended soil particles. Use two filters, and carefully increase the vacuum just until the sample begins to flow into the bottom half of the filter apparatus. If the filtrate still has floating particles, filter it a second time before proceeding to measure the nutrients. Transfer the clear filtrate into a clean 250-mL beaker, and use this solution to measure the nutrients.
5. Measure NPK using YSI kits. Your TA will tell you which dilutions are likely needed to bring the concentrations into the range that the photometer can measure. However, there can be variation based on when and where the sample was collected, so you might need to repeat a measurement using a different dilution. Specific instructions for each nutrient are posted on Blackboard and are available in the lab. Read Appendix 3 for more information about how to calculate concentrations using dilution factors and how to properly clean the glassware once you are finished.

Ask your TA for help as needed.

Soil permeability and field capacity: Soil permeability and field capacity are both affected the porosity of the soil, and thus the soil texture. Soil permeability measure how quickly water moves through the soil. This is very similar to water infiltration, which is the movement of water into the soil, such as after a rain or irrigation. The amount of water retained by the soil after drainage has occurred is the field capacity. We will measure permeability by timing how long a set volume of water takes to move through a soil sample. We will measure the field capacity by determining the difference between the amount of water added and the amount of water that passes through the soil sample.

1. Get a 500-mL water bottle that had the bottom removed. Cover the neck with mesh and secure with a rubber band. Turn the bottle upside down (neck down, opened bottom at the top, so it looks like a funnel), and place the bottle over a beaker.

2. Add 50 g of dried soil to the bottle. Measure 50 mL of water into a graduated cylinder. Start a timer as soon as you start pouring the water into the bottle. Stop timing once there is no more water draining from the bottle into the beaker. That time is the measurement of soil permeability.

Will highly permeable soils take more or less time to drain than low permeability soils?

_____

Measure the volume of water that collected in the beaker below the bottle using the graduated cylinder. The field capacity is the difference between 50 mL and the amount in the beaker.

How does the field capacity relate to the ability of plants to grow in the soil?

_____

Soil texture (via differential settling rates): We are going to estimate the soil texture using the settling rates of the different sized particles. Larger particles like sand settle faster than smaller particles like silts. We will mix a soil sample with water, and then observe the relative thickness of the amount of material that has settled to the bottom. The sand layer will be at the bottom, followed by the silt layer then the clay layer. Sand will settle within 1–2 minutes and silt within 2–3 hours, but clay can take several days or longer to settle! We will start the observations today, and then examine the samples again next week to view the layers.

1. Get a 500-mL water bottle and mark a line 4 cm and one 12 cm from the bottom. Add soil to the 4-cm line, then fill to the 12-cm line with water. Cap the bottle and shake for at least 60 seconds, until all the particles are suspended in the solution. Start a timer as soon as you set the bottle down. Allow it to stand undisturbed for 60 seconds, then mark the depth of the sand that has settled at the bottom.
2. If time allows, wait an additional 2 hours, then mark the depth of the silt that has settled above the sand. Clay will take several days to settle, and we will observe that next week. If a researcher wanted to determine the soil texture more accurately than the feel-test flowchart that we used, they could measure the thickness of each layer and converting them to percentages. They could then use the soil texture triangle (Fig. 11.4) to determine the texture class.

This demonstration relates to the texture of soils found in river floodplains. When there is a fast flow rate, the river can transport small and large particles, sometimes including large boulders. If the river overtops its channel and water flows into the floodplain, the flow rate decreases as the water spreads out over a larger area. As you observed in this demonstration, the larger particles are the first to settle out, while the smaller particles are carried throughout the floodplain. Eventually, those silt- and clay-sized particles will settle, or the water will evaporate, causing the amount of silt and clay to build up in the floodplains. These finer particles often have chemicals including nitrate and phosphate bound to them. The deposit of silt and clay in floodplains is one reason that floodplain soils tend to be more fertile than the surrounding uplands. This is reflected in our region in the higher concentration of agriculture in the Upper and Lower Valleys (this also is based on the increased availability of water in these parts of El Paso).

Did you find the expected differences when you compare the results for texture and nutrients of the desert mesa and river floodplain soils tested in this lab?

_____

Clean up:

1. Make sure not to pour any soil into the sink. We do not want to clog it! Rinse soil into the bucket next to the sink. Any soil that was not used can be returned to the appropriate planter on the Green Roof.
2. The NPK test solutions need to be disposed into the labeled waste container. Glass vials and plastic nitratest tubes need to be rinsed four times with DI water. Do not worry about getting all of the powder out of the nitratest tube.
3. The soil samples for the differential setting rate demonstrations can be disposed of in the regular garbage can.
4. Use the 95% ethanol to remove any writing from the glassware.
5. All of the glassware (including that made from plastic) needs to be washed with glassware soap, rinsed four times with tap water and four times with DI water. This includes the sieves, funnels, plastic spoons, the centrifuge tubes, the graduated cylinders, and the beakers. See Appendix 3 for more detailed instructions.

Assessments

Answer the questions in the lab assignment (available on Blackboard).

References

Bot, A., and Benites, J. 2005. _The Importance of Soil Organic Matter: Key to drought-Resistant Soil and Sustained Food and Production. FAO Soils Bulletin 80._ Rome, Italy: Food and Agriculture Organization of the United Nations.

United States Department of Agriculture (USDA). n.d. "Soil Formation and Classification." Accessed July 15, 2016. www.nrcs.usda.gov/wps/portal/nrcs/detail/soils/edu/?cid=nrcs142p2_054278.

USDA. 1999. "Soil Taxonomy: A Basic System of Soil Classification for Making and Interpreting Soil Surveys." www.nrcs.usda.gov/Internet/FSE_DOCUMENTS/nrcs142p2_051232.pdf. Accessed July 15, 2016.

Resources

An interview from 2012 with John Crawford about agriculture and soils: http://world.time.com/2012/12/14/what-if-the-worlds-soil-runs-out/

The Natural Resources Conservation Services has conducted soil surveys throughout much of the United States, and you can access their Web Soil Survey information at http://websoilsurvey.sc.egov.usda.gov/App/HomePage.htm. For example, much of the east side of El Paso is the Hueco-Wink association, which is a loamy sand in the aridisol order. Visit the webpage and find your neighborhood. What is the name and characteristics of the soil?

Introduction to Energy and Climate Change

Learning Objectives

1. Understand the causes and consequences of climate change
2. Relate personal choices about transportation to climate change
3. Investigate the economic, political, societal, and environmental costs and benefits of various energy sources

Importance

Climate change is one of the greatest environmental threats humanity faces. Global average temperatures are increasing, precipitation rates are changing, and sea level is rising. These changes are expected to reduce food production, increase the spread of disease transmitting insects, lead to increased wild fires and coastal flooding among a suite of other impacts. The main cause of climate change is the increase in the concentration of carbon dioxide in the atmosphere. Increased carbon dioxide traps more heat in the Earth's atmosphere, and this heat drives the other changes in the climate. Most of the carbon dioxide released by humans comes from the burning of fossil fuels, our main source of energy. Transitioning to low-or no-carbon energy sources will be critical in minimizing the amount of climate change we will experience in our lifetimes.

Introduction

The global average temperature has increased almost 1°C (1.8°F) over the past century (IPCC 2014). The increased temperatures are shifting precipitation patterns, with some regions receiving more rainfall while others, such as the American Southwest, are receiving less. Warmer temperatures also result in more precipitation falling as rain than as snow, resulting in smaller snowpacks. Warmer temperatures cause more glaciers to melt, which combined with the expansion of water as it warms, contributing to increased sea level. Higher sea levels increase flooding, and cause severe storms like hurricanes have even larger impacts due to flooding. You can read more about impacts of current climate change and possible consequences in the future in your textbook.

Although climate is influenced by many factors, the recent increase in temperature results from more carbon dioxide in the atmosphere. Carbon dioxide is a greenhouse gas, and acts to trap heat in the atmosphere. There are other greenhouse gases, but we will focus on carbon dioxide because it is the human-produced gas that currently has the largest impact. Human activities have caused carbon dioxide to increase to levels not seen in the past 800,000 years (IPCC 2014). Although the increase has been small in terms of absolute amounts, the small shift in the Earth's energy balance has caused warmer temperatures. Most of the carbon dioxide is released during the combustion of fossil fuels (coal, oil, and natural gas). Coal and natural gas are burned to generated electricity and for industrial processes, while oil is the main fuel for transportation. Increasing industrialization and development, coupled with rapid population growth, have caused the carbon dioxide concentrations to increase from 280 ppm in pre-industrial times to over 400 ppm today (IPCC 2014).

Low-or no-carbon energy sources are necessary to reduce and eventually stop the addition of carbon dioxide to the environment. Options such as wind, solar, nuclear, and fossil fuels with carbon capture and sequestration can provide electricity with a lower footprint than traditional fossil fuel burning. Electric vehicles that use batteries to store some of that carbon-free electricity can replace the use of oil. Transportation could also use biofuels, such as ethanol and biodiesel. You will learn about how to grow algae for producing biodiesel in the Biofuels lab in the population unit. Fossil fuels represent over 80% of the energy consumed globally (IEA 2015), so making this transition will not be easy.

In the first lab in this unit you will learn more about climate change, how transportation contributes to climate change, and which types of vehicles have the largest carbon dioxide emissions. In the second lab you will learn about the costs and benefits of different electricity-generating fuels. We will role-play the decision making process that El Paso Electric might go through while considering building a new power plant on the former American Smelting and Refining Company (ASARCO) site near UTEP. Each group will be assigned an energy source, and you will research the economic, political, societal, and environmental issues associated with that particular source. Each group will prepare a white paper trying to persuade the El Paso Electric Board of Directors to select their energy source. The white paper will be summarized during in-class presentations, followed by a discussion among the groups to further explore the costs and benefits of each source. At the end of this unit, you will understand the causes and consequences of climate change and some of the challenges and opportunities of switching to low-or no-carbon energy sources.

References

Intergovernmental Panel on Climate Change (IPCC). 2014. "*Climate Change 2014: Synthesis Report. Contribution of Working Groups I, II and III to the Fifth Assessment Report of the Intergovernmental Panel on Climate Change*" (Core Writing Team, R. K. Pachauri and L. A. Myers [eds.]). Geneva, Switzerland: IPCC, 151. IPCC AR5 Synthesis Report website: http://ar5-syr.ipcc.ch/. Acceded May 16, 2016.

International Energy Agency (IEA). 2015. "*Key World Energy Statistics 2015.*" http://www.iea.org/publications/freepublications/publication/key-world-energy-statistics-2015.html. Accessed May 16, 2016.

Lab 12
Vehicles and Climate Change

Learning Objectives

1. Understand processes that contribute to the greenhouse effect
2. Describe how humans are impacting the climate
3. Investigate the relationship between vehicle type and carbon dioxide emission
4. Understand trade-offs in sampling strategies
5. Use Excel for basic calculations and constructing graphs

Importance

Human activities, mostly burning of fossil fuels, have increased the concentration of greenhouse gases to levels not seen in thousands of years. These greenhouse gases trap heat in the atmosphere, contributing to a temperature increase of almost 1°C since 1900. The changing climate is already causing observable changes in temperature and precipitation patterns globally. One significant source of carbon dioxide, a major greenhouse gas, is transportation. Vehicles differ in their levels of fuel consumption, producing greater or lesser amounts of carbon emissions during combustion of gasoline. Reducing transportation's contribution to climate change will require driving less, shifting vehicles to low-or no-carbon fuels, and improving the fuel economy of all vehicle types.

Guiding Questions

How much of your ecological footprint comes from transportation?

How does vehicle choice relate to climate change?

Should the amount of carbon emission influence someone's vehicle buying decision?

Are personal vehicles an efficient means of transportation?

What alternative choices could you make to reduce the carbon emissions resulting from your transportation needs?

What assumptions are made when extrapolating from a sample to an entire population?

How can graphs help us understand patterns and interpret results of experiments?

Adapted from *Environmental Science* by Wagner and Sanford (2009).

Vocabulary

1. *Greenhouse gases*—Atmospheric gases that absorb heat and reradiate some back to the surface of the planet, increasing its average temperature. The major greenhouse gases are water vapor, carbon dioxide, methane, and nitrous oxide.
2. *Fossil fuels*—Coal, oil, and natural gas. They formed from organic materials deposited millions of years ago.
3. *Fuel economy*—The amount of fuel required for a vehicle to travel a certain distance, given as miles per gallon (mpg) in the United States.
4. *Corporate average fuel economy (CAFE) standards*—Regulations that set the average fuel economy of a manufacturer's fleet of vehicles for a given model year. The standards are designed to push manufacturers to produce vehicles with better fuel efficiency.
5. *Population (statistical)*—All of the items or organisms that have the characteristic of interest. In this lab, the population is all of the vehicles in the United States.
6. *Sample (statistical)*—One or more observations from the population. In this lab, the sample will be 25 vehicles on the UTEP campus.
7. *Census (statistical)*—The collection of data from all members of the population.
8. *Bias (statistical)*—The difference between the sampled data and the population data. Biased data can be collected when sampling is not well designed.

Introduction

Climate change and transportation

Over the past 200 years, human activities including the burning of fossil fuels (e.g., coal, oil, natural gas) have increased the concentration of heat-trapping "greenhouse gases" in the Earth's atmosphere. Greenhouse gases are necessary for life as we know it, as they keep the Earth's temperature warmer than it would otherwise be. But the anthropogenic (human-caused) increases of greenhouse gases (primarily carbon dioxide–CO_2–and methane–CH_4) have largely caused an increase of about 0.8°C (1.4°F) in the average global temperature over the past 100 years (IPCC 2014). As the concentration of greenhouses gases increase in the atmosphere, less heat can escape to outer space. Other aspects of the climate are also changing, such as rainfall and winds patterns. Related events include changes in snow and ice cover and in sea level. If greenhouse gas emissions continue to increase, climate models predict that the Earth's average surface temperature could increase from 0.3°C to 4.8°C (0.5°F to 8.6°F) above 1990 levels by 2100 (Fig. 12.1; IPCC 2014). Much of the variation in the estimates depends on what actions people take. If there are significant reductions in greenhouse gas emissions, the temperature increase will be at the lower end of the range (maps on the left side of Fig. 12.1). If we continue with business as usual, the increase will be at the upper end (maps on the right side of Fig. 12.1). You can read more about the evidence for and consequences of climate change, and actions that can impact future climate change, in the Global Climate Change chapter of your textbook.

In the United States, fossil fuel combustion accounts for about 77% of our anthropogenic greenhouse gas emissions, mostly in the form of carbon dioxide emissions (US EPA 2015a).

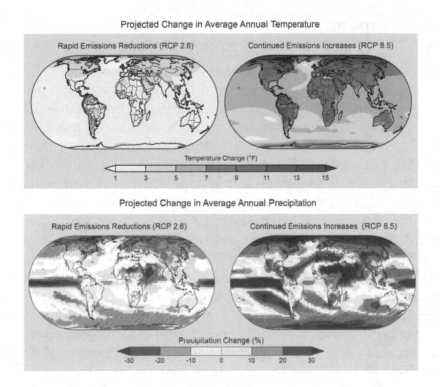

Figure 12.1. Projected changes in surface temperature (top) and average precipitation (bottom) for 2081–2100. The maps on the left (RCP 2.6) are based on the assumption that greenhouse gas emissions peak by 2020, and rapidly decline afterward. The maps on the right (RCP 8.5) assume that greenhouse gas emissions continue to rise throughout the century, reaching 1200 ppm by 2100. The current concentration of greenhouse gases is around 400 ppm. Figures from NOAA NCDC/CICS-NC, part of the Climate Change Impacts in the United States, available at http://nca2014.globalchange.gov/.

Carbon dioxide is the product of complete combustion of fuel. Carbon dioxide does not directly impair human health unless in concentrations much higher than currently exist, but is regulated by the U.S. Environmental Protection Agency (EPA) because climate change threatens the "health and welfare of current and future generations" (US EPA n.d.). There are currently policies that aim to reduce carbon emissions from power plants (announced in 2015, US EPA 2015b) and from vehicles (started in 2012, with standards set to 2025 model year vehicles, US EPA 2012).

In 2013, transportation activities contributed 27% of the total US greenhouse gas emissions, second only to electricity generation (US EPA 2015a). Passenger cars accounted for 43% of the transportation greenhouse gases, while light duty trucks (pickup trucks, SUVs) accounted for 17% of the total (US EPA 2015a). The combustion of a typical gallon of gasoline releases 8.8 kg (19.4 lb) of CO_2 (Wagner and Sanford 2009). Conventional vehicles are not fuel efficient, with only about 14% of the energy in the gasoline actually moving the vehicle forward, with the rest lost as engine heat, to friction, and to idling (Withgott and Laposota, 2015). As part of the CO_2 regulations on vehicles, the CAFE standards are required to reach 54.5 mpg by 2025, compared with 27.5 mpg from 1990–2010 (http://www.nhtsa.gov/fuel-economy). You can read more about fuel efficiency in the Nonrenewable Energy Sources chapter in the textbook.

Estimation and sampling

The most accurate way to determine the total annual CO_2 emissions from all the vehicles in the US would be to go to every vehicle, install a device that measures CO_2 emissions, and return in a year to retrieve the results. However, this type of survey is not a feasible option in a country with over 300 million people, and is probably not even feasible in a single city such as El Paso. Instead, we use sampling methods that allow reliable estimates of CO_2 emissions. Sampling requires us to make assumptions including that the sample accurately represents the entire population of interest and is unbiased. Random sampling is a robust method that requires that all individuals in the population have an equal opportunity to be sampled (Fig. 12.2). A truly random sample requires the use of a random number generator or something similar to determine which individuals in the population are actually sampled. Because truly random samples can be difficult to obtain, systematic sampling is sometimes used (Fig. 12.3). In this method, samples are collected at regular intervals, such as every third vehicle or every 10th student. Systematic sampling can give biased results if there are underlying patterns in the data, such as large vehicles tending to be parked away from each other because the drivers prefer more space. Defining a population and determining how to sample that population is a critical part of research and is a complex topic, most of which is outside the scope of this class. However, you should be asking yourself if samples taken are *representative, random,* and *unbiased* for any study that you conduct or read about. During this lab, you should ask yourself: Are the vehicles in one parking lot representative of all of UTEP? Are the vehicles at UTEP representative of the entire United States? Is it appropriate to extrapolate from one sample to the entire United State? How do we balance the time (and money) required to conduct a study with the need to obtain accurate and unbiased results?

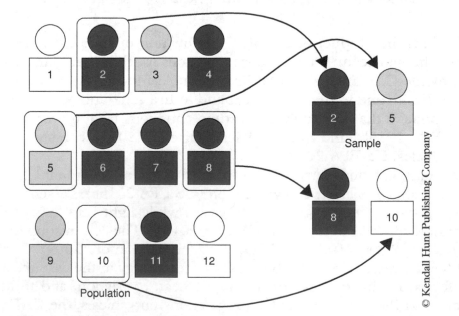

© Kendall Hunt Publishing Company

Figure 12.2. An example of a simple random sample.

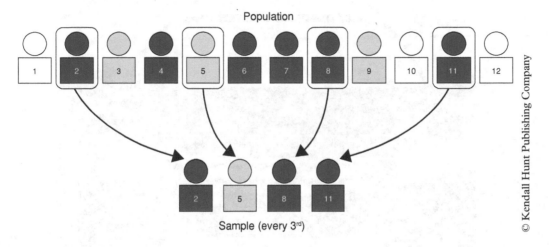

Sample (every 3rd)

Figure 12.3. An example of systematic sampling.

Activities

Today we will be collecting basic data and doing calculations to estimate CO_2 emissions and fuel consumption of an average vehicle on the UTEP campus, and then using those values to estimate the total CO_2 emissions and fuel consumption of vehicles in the entire United States. Each group will be assigned a parking lot where you will record the make, model, and year of every third vehicle present (a systematic sampling design). You will then look up the fuel consumption and CO_2 emissions of these vehicles using the EPA's website, and use the average values to extrapolate to the entire United States. We will also compare the fuel consumption of several vehicle types.

Sampling vehicles

1. Each group will be assigned a UTEP parking lot. At the parking lot, record **every third** vehicle's make, model, and year (e.g., Toyota Camry, 2013). If possible, also record information such as 2WD or 4WD, V6 or V8, automatic or manual transmission, and so forth, as these are factors that influence fuel economy. If you do not know the exact year or model, you can use the VIN number to determine the year of the vehicle. The VIN is a 17-digit number that is visible in the windshield of all vehicles (Fig. 12.4), and the 10th digit is the code for the year the vehicle was made (Table 12.1). ***When collecting data, do not lean on or set anything on the vehicles!*** If anyone asks you what you are doing, politely inform that you are a student in the Environmental Science lab and are only collecting information about their vehicle for a class project. Although very few people have asked questions in the past, those who do are usually worried that you are giving them a parking ticket! You can always skip a vehicle if its owner does not want you collecting any data.
2. Collect data for 25 vehicles (cars, trucks, SUVs, and vans), **using every third vehicle**. Use the *Parking Lot Tally Sheet* to record your data.

Victor Metelskiy/Shutterstock.com

Figure 12.4. Location of VIN on all vehicles.

Table 12.1: The VIN code for the year a vehicle was made

The 10th digit (or eighth from the right) is the code of the year. For example, VIN WAULFAFR2**B**A086039 is from 2011, while 3B7HC13Y1**W**G231043 is from 1998.

Digit	Year	Digit	Year	Digit	Year	Digit	Year	Digit	Year
A	1980	L	1990	Y	2000	A	2010	L	2020
B	1981	M	1991	1	2001	B	2011	M	2021
C	1982	N	1992	2	2002	C	2012	N	2022
D	1983	P	1993	3	2003	D	2013	P	2023
E	1984	R	1994	4	2004	E	2014	R	2024
F	1985	S	1995	5	2005	F	2015	S	2025
G	1986	T	1996	6	2006	G	2016	T	2026
H	1987	V	1997	7	2007	H	2017	V	2027
J	1988	W	1998	8	2008	J	2018	W	2028
K	1989	X	1999	9	2009	K	2019	X	2029

Table by K. Floyd with data from National Highway Traffic Safety Administration.

Estimating fuel economy and carbon emissions

1. Go to the US Environmental Protection Agency's Fuel Economy website at www. fueleconomy.gov/feg/findacar.htm.

2. Search for each vehicle and record the following on the Parking Lot Tally Sheet:

   - combined estimated fuel consumption (barrels per year: 1 barrel = 42 gallons; use Excel to perform the calculation using a formula, and then copy and paste the formula to do the calculations for all 25 vehicles)
   - annual tailpipe CO_2 emissions (metric tons per year)
   - cost to drive 25 miles
   - annual fuel costs (assume 15,000 miles driven per year)

   Note that the information is on two separate tabs on the website: "Fuel economy" and "Energy and Environment." You will also need to change the units for greenhouse gas emissions from "grams per mile" to "metric tons per year."

3. Input the data into a Excel spreadsheet (posted on Blackboard). As a group, calculate a parking lot *sample mean* and *standard deviation* for annual fuel consumption (in gallons) and annual CO_2 emissions for:

   - All vehicles
   - Cars only
   - SUVs only (include vans with SUVs)
   - Pickup trucks only

   If you are familiar with Excel, this is easy to do by sorting the data and using formulas. If you are not familiar, there are instructions posted on Blackboard and your TA will help with any questions.

4. Make a bar graph of the *annual CO_2 emissions* comparing cars only, SUVs only, trucks only, and all vehicles together. This should be done in Excel, see instructions on Blackboard.

5. Use the mean values obtained in step 3 to calculate total annual fuel consumption and CO_2 emissions for the entire *United States*, assuming there are 234,000,000 vehicles. Use the values for **all vehicles**, without separating out cars, SUVs, and pickup trucks for this step.

Assessments

Copy the graph into a Word document to print. Attach it to the lab assignment. Although the data and graph will be the same for all of the students in your group, you need to answer the analysis questions individually.

References

IPCC. 2014. Climate Change 2014: *Synthesis Report. Contribution of Working Groups I, II and III to the Fifth Assessment Report of the Intergovernmental Panel on Climate Change* (Core Writing Team, R. K. Pachauri and L. A. Myers [eds.]). Geneva, Switzerland: IPCC, 151. IPCC AR5 Synthesis Report website: http://ar5-syr.ipcc.ch/.

U.S. Environmental Protection Agency (US EPA). 2012. *FACT SHEET: EPA and NHTSA Set to Reduce Greenhouse Gases and Improve Fuel Economy for Model Years 2017–2025 Cars and Light Trucks.* www3.epa.gov/otaq/climate/documents/420f12051.pdf.

US EPA. 2015a. *Inventory of U.S. Greenhouse Gas Emissions and Sinks: 1990–2013.* www3.epa.gov/climatechange/ghgemissions/usinventoryreport.html.

US EPA. 2015b. *FACT SHEET: Overview of the Clean Power Plan.* Accessed December 9, 2015. www.epa.gov/cleanpowerplan/fact-sheet-overview-clean-power-plan.

US EPA. n.d. *Endangerment and Cause or Contribute Findings for Greenhouse Gases under Section 202(a) of the Clean Air Act.*, Accessed December 9, 2015. www3.epa.gov/climatechange/endangerment/index.html.

Wagner, T., and Sanford, R. 2009. *Environmental Science: Active Learning Laboratories and Applied Problem Sets.* 2nd ed. London: John Wiley and Sons.

Resources

Examples of statistical samples and populations: http://courses.tlt.psu.edu/course/bio12/module03/2009/10/lesson-02-samples-and-populations.html.

Short (4 min) video from Neil deGrasse Tyson explaining the greenhouse effect and how human activity is affecting it: www.youtube.com/watch?v=6VUPIX7yEOM

Tips from the EPA on improving gas mileage: www.fueleconomy.gov/feg/drive.shtml

Lab 13
Energy Development Scenarios

Learning Objectives

1. Increase your energy literacy: "[A]n understanding of the nature and role of energy in the world and daily lives accompanied by the ability to apply this understanding to answer questions and solve problems."—http://energy.gov/eere/education/energy-literacy-essential-principles-and-fundamental-concepts-energy-education

2. Learn about the economic, political, societal, and environmental costs and benefits of various energy sources

3. Develop persuasive arguments in both written and spoken forms

4. Practice critical evaluation of information available on websites, to distinguish between fact and opinion

Importance

Much of modern life is driven by our consumption of energy, from electricity used to watch TV and charge cell phones to gasoline or diesel used to transport people and goods worldwide. As the human population continues to increase from over 7 billion people in 2015 to projections of 10 billion by the end of the century, along with increasing development in many countries, our use of energy is expected to greatly increase. Currently, most of our energy demand is met by fossil fuels: coal, natural gas, and oil. Using these fuels has significant environmental, economic, and health impacts, including habitat destruction during the extraction of the fuels, air and water pollution, and changing the global climate during combustion. Shifting to other energy sources, such as nuclear, hydropower, wind, or solar, can reduce many of the problems associated with fossil fuels, but increased use of these sources also have costs. The speed at which countries can make the shift away from fossil fuels will determine how severe the consequences of climate change, along with the other costs of burning fossil fuels, will be.

Guiding Questions

How would you weigh trade-offs of availability, cost, and environmental impacts in deciding which type of energy source should be used?

How would you balance short-term costs and benefits against long-term costs and benefits for each energy source?

Vocabulary

1. *Electricity*—The presence and flow of electric charge. It is convenient to use because it can be transmitted over large distances through cables and can be used in many ways (lighting, heating, cooking, industrial equipment, etc.).
2. *Watt* (W)—The metric unit of power. Electricity is generally measured in terms of kilowatts. (kW, 1000 W) at the local scale and megawatts (MW, 10_6 W) or gigawatts (GW, 10_9 W) at state, national, and global scales.
3. *Kilowatt-hour* (kWh)—The measure of electricity consumption, amount of energy equivalent to one kilowatt of power expended for 1 hour.
4. *Million tons of oil equivalent* (Mtoe) —The energy released by burning 1 million tons of crude oil. Used in some reports to standardize different energy sources into one unit for comparison purposes.
5. *Electrical grid*—The interconnected network of generating stations, high-voltage transmission lines, and distribution lines that connects suppliers to consumers.
6. *Renewable energy sources*—Solar, wind, hydroelectric, biomass, tidal. These sources are either renewed on a human timescale (hydroelectric and biomass) or are not consumed in such a way that reduces the supply (solar, wind, tidal). Their supply is effectively unlimited and generally there are no fuel costs.
7. *Nuclear power*—Uses radioactive decay of uranium (isotope U-235) to heat water, turning it to steam. The steam spins a turbine, generating electricity. Although there is little carbon pollution, the supply of uranium is finite, so nuclear power is not a renewable energy source. Radioactive uranium fuel and waste products can potentially cause increased incidence of cancer and other health risks if not properly used and stored.
8. *Greenhouse gases* (GHG)—Atmospheric gases that absorb and emit thermal infrared radiation (heat). The main GHG are carbon dioxide (CO_2), methane (CH_4), nitrous oxide (N_2O), and water vapor (H_2O). Without GHG the average temperature of the Earth would be about −18°C (0°F). With the greenhouse effect caused by GHG trapping some of the heat radiating from the Earth, the average temperature is instead around 14°C (58°F).
9. *Climate change*—Increase in the average global temperature caused mostly by an increase in anthropogenic (human-caused) GHG. Since 1900, the average temperature has increased by 0.7°C–0.8°C (1.2°F–1.4°F). Increased temperatures cause changes in the amounts and timing of precipitation, increased melting of glaciers and ice caps, and sea level rise, among other effects.

Introduction

Global patterns of energy use

Global energy use has increased about 45% in the past 20 years (from approximately 9,000 to 13,000 million tons of oil equivalent [Mtoe]), and is projected to increase an additional 45% between 2010 and 2035 (Tollefson and Monastersky 2012). These increases carry a 46% increase in carbon emissions. Carbon emissions contribute to anthropogenic climate change (review the Climate Change and Vehicles lab in this unit and the Global Climate Change chapter of your textbook for more information). To keep atmospheric CO_2 levels below 450 ppm (an amount thought to keep the global average temperature increase to

2°C), we can either decrease the total amount of energy used globally or we can switch from carbon-intensive fuels to those which produce fewer carbon emissions. However, all forms of energy have some form of environmental impact, especially when scaled to amounts large enough to meet the demands of billions of people across the globe. In addition, each source can have economic, political, and societal costs and benefits.

This lab will focus on our use of electricity. All electricity is generated using other naturally occurring fuels, such as fossil fuels, radioactive elements (nuclear), wind, solar, or biofuels. These are primary energy sources. Primary energy sources also include petroleum for transportation and natural gas and coal used directly for industrial processes. About 25% of the total primary energy production is used to produce electricity (International Energy Statistics, USEIA).

It is projected that global electricity capacity needs will more than double by 2040 to both meet increasing global demand and to replace power plants scheduled to retire (World Energy Outlook 2014 Executive Summary, IEA). The total electricity capacity in 2012 was 5,500 GW, and total electricity consumption was almost 19.4 trillion kWh (International Energy Statistics, USEIA). About 67% of the electricity was generated with fossil fuels, about 10% with nuclear fuel, and 22% with renewable energy sources (International Energy Statistics, USEIA). Hydroelectric is the main source of renewable electricity, at 77%, followed by wind at 11%, biomass and waste at 8%, and solar at 2% (International Energy Statistics, USEIA). Renewables are projected to increase rapidly, with wind and solar expected to quadruple by 2040 (World Energy Outlook 2014 Executive Summary, IEA).

United States patterns of energy use

In 2013, the United States generated approximately 4 trillion kWh of electricity, about 20% of the global total (Figure 13.2; Annual Energy Outlook 2015, USEIA). Burning of fossil fuels contributed 67% (27% natural gas, 39% coal, and 1% petroleum), nuclear contributed 19%, and renewables contributed 13% to the total (Fig. 13.1). This represented a large increase in natural gas usage since 2000 and a large decrease in the use of coal (Fig. 13.1). About half of the renewable electricity is produced by hydropower, but wind and solar have increased rapidly over the past several years (Fig. 13.2; Annual Energy Outlook 2015, USEIA). The US Energy Information Agency projects an increase in total electricity production to 5 trillion kWh by 2040, with most of the increase coming from increased use of natural gas and renewables (Fig. 13.1). These projections can change depending on costs of fuels or equipment like photovoltaic panels or wind turbines, and based on changes in laws and regulations. For example, as new regulations of CO_2 emissions begin, such as the Environmental Protection Agency's (EPA) Clean Power Plan to reduce carbon emissions from power plants that was announced in August 2015, burning fossil fuels to generate electricity will become more expensive, which might shift more of the electricity generation to low-carbon sources such as renewables or nuclear.

El paso patterns of energy use

El Paso Electric (EPE) supplies electricity to almost 400,000 customers in west Texas and southern New Mexico (EPE 2014). Their generating capacity of more than 1,800 MW in 2014 consisted of approximately 47% nuclear fuel, 35% natural gas, 5% coal, 13% purchased power, and <1% of company-owned solar photovoltaic panels and wind turbines (EPE 2014). The purchased power includes 107 MW from five solar photovoltaic generation facilities (EPE 2014). In 2016, the company announced that they sold their ownership in the Four Corners coal-fired power plant and that they are now coal-free (EPE 2016).

It is projected that El Paso Country will grow from around 800,000 people in 2010 to nearly 1.3 million people in 2050, an increase of 37% (2014 Data Projection Tool, Office of the State Demographer). EPE has estimated that it must spend approximately $1.1 billion during 2015–2019 in construction projects to meet increasing load requirements and to replace older plants and terminated purchased power agreements (EPE 2014). Currently,

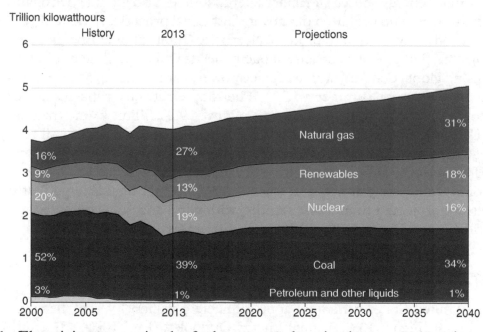

Figure 13.1. Electricity generation by fuel source and projections to 2040 using a reference case (assumes moderate economic growth, no major changes in current laws and policies). This is Figure 31 in the Annual Energy Outlook 2015, USEIA.

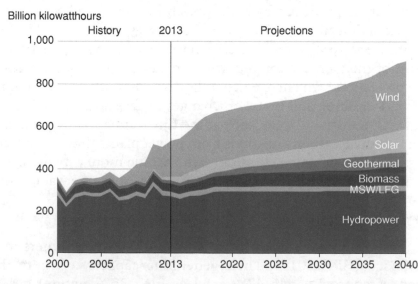

Figure 13.2. Renewable electricity generation by fuel source and projections to 2040 using a reference case (assumes moderate economic growth, no major changes in current laws and policies). Wind and solar are projected to increase the most during this period. This is Figure 34 in the Annual Energy Outlook 2015, USEIA.

about half of those funds have been allocated for building a new natural gas burning power plant in east El Paso.

Scenario for discussion

Let's assume that EPE is considering using the former American Smelting and Refining Company (ASARCO) site near University of Texas at El Paso (UTEP) (Fig. 13.3) for building a new power plant to help meet the projected demand for electricity. They are considering solar, wind, natural gas, coal, or nuclear options. The Board of Directors is taking comments from representatives of companies selling each energy option before making their decision. The plan is to start the construction of the power plant within the next 5 years, so comments need to address current technologies (e.g., those possible in the next year or two).

Activities

Your group will be assigned an energy source to investigate.

Before this week's lab, each group will need to prepare a written white paper (1–2 pages) and a short (5–7 minutes) presentation (see section Prior to Coming to Lab). You will have 5–10 minutes at the start of lab to make final preparations for the presentation, but most of the work needs to be completed before coming to lab. Each group will make their presentations to the EPE Board of Directors (your TA), followed by time for open discussion. During the discussion, each group will ask questions of the other groups, trying to find out more information about their position or trying to highlight problems with positions and/or arguments of others. Each group needs to consider the arguments for and against their

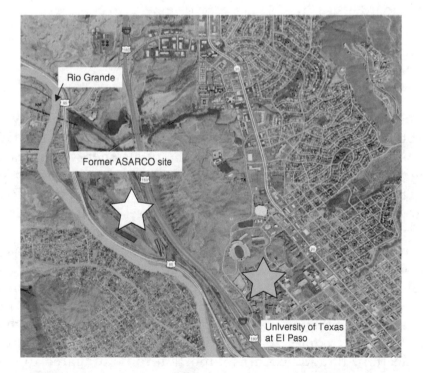

Figure 13.3. Location of the former ASARCO site and its relation to UTEP and the Rio Grande. Map by K. Floyd with imagery from US Geological Survey EROS Orthoimagery (https://raster.nationalmap.gov/arcgis/rest/services/Orthoimagery).

position, as the arguments against are likely to be brought up during the open discussion. At the end of the lab, there will be a final assessment including a vote on which energy source EPE should install.

Some issues to consider when crafting your arguments: cost of electricity, cost of health care, reliability of energy throughout the day and year, future cost increases due to changes in fuel costs, future costs due to changes in regulations, use of water in generating electricity, waste disposal, consequences of accidents, job creation, quality of life, benefits to stakeholders. This is not an exhaustive list, but rather some of the major issues that arise when thinking about options for energy production choices. There is a short list of the pros and cons for each energy sources posted on Blackboard. You should also review the energy chapters in your textbook. You might also find it helpful to review the climate change and air pollution chapters.

Be careful when researching your position online. Energy use and its impact on the environment are hugely important in the United States (and the world), and there are lots of websites that present biased opinions as facts. You must critically evaluate the information on each site, and should check multiple websites to assess the validity of information. Always try to find an "about us" section on a webpage to learn about the particular point of view that is likely being expressed. There are also several websites listed in the resources section that we think are good places to begin your research.

Assessments

There are two major components to this lab: the white paper (a separate grade) and the presentation and discussion (the lab assignment for this week). Grading rubrics for each are provided on Blackboard.

Prior to coming to lab

White paper

White papers generally are persuasive tools, arguing a point of view or proposing a solution to a problem. For this assignment, you will be writing to the EPE Board arguing in favor of your energy source and against the other energy sources. General guidelines for writing white papers can be found at website listed in the resources section. The white paper begins with *background* focused on the problem that EPE needs to solve, namely producing more electricity. Next, the paper presents the **solution to the problem**. In this section, it is necessary not only to describe why your energy source is the best option, but to also describe why the other energy sources are not as good. You need to consider evironmental, economic, political, and societal costs and benefits. Because this is a persuasive document, you can focus on the benefits of your energy source and the costs (negative aspects) of the other energy sources. A strong argument is aware of the strengths and weaknesses of a position, and can address the strengths while downplaying the weaknesses. However, completely ignoring weaknesses puts the credibility of your argument into question. In addition, it is important to be prepared to counter the likely arguments raised against your energy source by other groups during your presentations and discussion. Finally, write a **conclusion** that summarizes your position. Each white paper must include a **literature cited** section for all of the resources that you used, and you need to include **in-text citations** in the body of the white paper wherever you used outside information (including the textbook).

White paper outline

Introduction:

- What is the problem you are going to solve?

Solution to the problem:

- Briefly state your solution.
- Explain the benefits of your solution, including economic, political, societal, and environmental issues (as relevant).
- Explain why your solution is better than the alternatives (think about the weaknesses of the other energy sources relative to yours, again considering including economic, political, societal, and environmental issues).
- Explain at least two potential weaknesses with your solution, and why they are not important enough for EP Electric to choose a different energy source (this could be combined with the previous bullet point if that makes your argument stronger). Note that this will help you be prepared to refute arguments from the other groups during the class discussion.

Conclusion

- Summarize your position in a clear and concise way.

Literature cited and references

- Include both in-text citations (parenthetical format: [Author, date]) and a final literature cited list. You can use any format that you are familiar with. Go to www.studygs.net/citation.htm for instructions on how to cite web pages.

Presentation

Prepare a *5–7 minute oral presentation* based on your white paper. Be sure to make a strong case why the EPE Board of Directors should follow your position. You will be assessed on:

- Vocal and physical delivery
- Knowledge of topic
- Covering the components mentioned in the white paper (see earlier section)
- Quality of argument

- Persuasiveness of argument

Be prepared to answer the following questions (from *Affordable Energy for Our Future*. National Geographic).

- What is your position?
- What economic, political, and/or social factors contribute to your viewpoint?
- What environmental and/or technical factors contribute to your viewpoint?
- How would you refute your opponent's viewpoint?
- How could some of your opponent's arguments support your own argument?
- Which of your arguments are technical and scientific, and which are based on values and beliefs?

In lab:

1. Give your oral presentation (assessed by peer reviews, rubric provided on Blackboard).
2. Listen carefully to the other presentations. If they present a position counter to yours, think about weaknesses in their argument that you can ask them to explain. If they have a position that supports yours, think about questions you can ask that highlight that support.
3. Open discussion forum where you can ask questions of each group.
4. At the end, assess which group (other than yours) presented the strongest argument in their opening statement and in their asking and answering of questions in the forum. The group that is **ranked the highest** by their peers will receive **five points of extra credit** on this assignment!
5. Turn in the white paper.

References

El Paso Electric (EPE). 2014. "Annual Report 2014." Accessed September 18, 2015. http://ir.epelectric.com/annuals.cfm.

EPE. 2016. "EPE becomes the First Regional Utility to Go Coal-Free." Accessed October 31, 2016. https://www.epelectric.com/about-el-paso-electric/article/epe-becomes-the-first-regional-utility-to-go-coal-free.

International Energy Agency (IEA). 2014. *World Energy Outlook 2014 Executive Report*. Accessed September 17, 2015. www.worldenergyoutlook.org/publications/weo-2014/.

National Geographic. n.d. "Affordable Energy for Our Future." Accessed September 15, 2015. http://education.nationalgeographic.com/activity/affordable-energy-our-future/.

Office of the State Demographer, Texas State Data Center. 2014. "Texas Population Projections by Migration Scenario Data Tool." Accessed September 17, 2015. http://osd.texas.gov/Data/TPEPP/Projections/Tool?fid=F211C21D04004A8EB499E976D712EA0B.

Tollefson, J., and Monastersky, R. 2012. The global energy challenge: Awash with carbon. *Nature* 491: 654–5. http://www.nature.com/news/the-global-energy-challenge-awash-with-carbon-1.11909.

U.S. Energy Information Administration (USEIA). 2015. *Annual Energy Outlook 2015*. Accessed September 15, 2015. http://www.eia.gov/forecasts/aeo/.

USEIA. n.d. "International Energy Statistics." Accessed September 18, 2015. http://www.eia.gov/cfapps/ipdbproject/IEDIndex3.cfm?tid=2&pid=2&aid=2.

Withgott, J. and Laposota, M. 2015. *Essential Environment: The Science behind the Stories*. 5th ed. San Francisco: Pearson Publishing Co.

Resources

What you need to know about energy. The National Academies of Sciences, Engineering, and Medicine. Uses data from the late 2000s, so some trends have changed, but overall easy-to-understand information about energy use, sources, costs, and energy efficiency. http://needtoknow.nas.edu/energy/.

Department of Energy website: http://energy.gov. Lots of information, can potentially be overwhelming at first. For some easy-to-digest information, check out their "Top things you didn't know about ..." series: http://energy.gov/joules-wisdom-top-things-you-didnt-know-about-energy.

The Great Energy Challenge. National Geographic. http://news.nationalgeographic.com/energy/. Lots of good articles about all aspects of energy use. National Geographic also has an online encyclopedia with detailed but pretty easy to understand information: http://education.nationalgeographic.com/encyclopedia/

The Wikipedia pages for energy sources are pretty good, although as usual be critical when evaluating the content. The page for overall electricity generation is also good for an introduction to the topics (https://en.wikipedia.org/wiki/Electricity_generation). The page for environmental impacts is good (https://en.wikipedia.org/wiki/Environmental_impact_of_the_energy_industry), and there is a page specifically about the environmental impacts of electricity generation (https://en.wikipedia.org/wiki/Environmental_impact_of_electricity_generation). These last three pages are particularly good for groups that have roles not directly associated with a particular energy source (e.g., health care providers or environmental activists).

More information about writing white papers can be found at the Purdue Online Writing Lab (https://owl.english.purdue.edu/owl/resource/546/1/) and in an article by Michael Stelzner (http://coe.winthrop.edu/educ651/readings/HowTo_WhitePaper.pdf).

Lab and River Safety

Basic Laboratory Safety

The laboratories planned for this semester are safe. However, we will be working with physical and chemical hazards that require safe handling procedures. Please take a few minutes to review these practices. In the future, as you progress in your scientific training, you may be faced with handling HAZARDOUS chemicals and substances. Therefore, it is essential that you develop good laboratory practices now. In general, if you are not sure of something, ask your instructor! It may prevent a serious problem.

Carelessness and *ignorance* are the most common cause of personal injury in the laboratory. It is essential that a student follow the instructions given by the instructor.

Good laboratory practices related to your personal safety

1. Eating, drinking, smoking, gum chewing, applying cosmetics, and taking medicine in the lab is NOT allowed. This includes having water bottles in the lab. All food and drink must be left on the table outside the lab.
2. You must wear appropriate clothing to the lab. You must wear closed-toe shoes. No sandals or similar footwear are allowed. If a chemical is spilled, your foot needs to be protected. You will *not be allowed in the lab* if you are not wearing closed-toe shoes, and will not be able to make up any missed work.
3. If you have long hair, tie it back so there is no chance it will get contaminated with a chemical or catch on fire. Also be aware of any dangling jewelry or loose clothing that might get contaminated.
4. Keep work areas clean. Do not pour any chemical down the sink. Dispose of chemicals only in labeled containers designated for disposal.
5. Do not taste or inhale any material. Work with chemicals in the chemical fume hood if appropriate (instructor will note when this is the case).
6. Familiarize yourself with the health and safety hazards of the equipment and chemicals with which you are working. Safety Data Sheets (SDS) are available for your review. An SDS describes potential hazards associated with working with a substance and emergency response procedures. The SDS also includes information about proper waste disposal. Examples are available on Blackboard.
7. Handle hazardous chemicals carefully. Do not move them around the room uncovered or without a secondary containment container. Place them toward the back or center of the bench so there is no chance they will be knocked over. Never return unused chemicals to the stock bottle. Do not use unlabeled chemicals and label all containers with their contents.
8. Familiarize yourself with the eyewash station and its use, fire extinguishers, first aid kits, and other emergency response equipment and exits.

9. Never pipette by mouth.
10. Wash your hands before leaving the lab. It is very easy to carry small amounts of potential noxious chemicals on your hands, and then proceed to eat lunch or dinner and ingest these compounds unknowingly.

Emergency response procedures

1. If you are injured, notify the instructor IMMEDIATELY.
2. If a chemical spill occurs, notify the instructor IMMEDIATELY.
3. If the evacuation alarm is activated, leave the building IMMEDIATELY with your instructor and await further instructions.
4. Emergency Contact Numbers:
    University Police   915-747-5611
    Department of Biological Sciences (BioSci 2.120)   915-747-5844
    Environmental Health and Safety   915-747-7124
    Life Threatening Situations   Dial 911
    Facility Emergency   915-747-7187

You can read the Laboratory Safety Manual provided by UTEP's Environmental Health and Safety Office for more information: http://ehs.utep.edu/manuals.html

River Safety

The field trip to the Rio Grande is always fun and informative. The water level is generally low during the winter, when water is stored in Elephant Butte Reservoir in New Mexico, and high during the spring and summer, when water is released for farmers to use for irrigating their crops. Although the river is not particularly deep, **it is a dangerous place, even at low flow**. Always use caution, and follow these safety instructions.

No one is required to go into the water. If you are not comfortable in water or around water, volunteer to take part in a different aspect of the lab.

1. Nonswimmers cannot enter the water under any circumstances.
2. Always have a buddy when in the water—someone on shore watching with rope or a life ring handy.
3. Always wear a life jacket in the water.
4. Always use a pole to check the bottom for stability before walking. Rivers carve out deep pockets that can't be seen from the surface, so don't assume that the river bottom is level. The depth can change dramatically even in one small step.
5. If you slip, remove the chest waders immediately. Try not to panic, roll to your back so that your head is above water, and look to the shore for the person throwing the rope or life ring.
6. In case of an emergency, **do not** enter the water yourself. Notify the TA immediately.
7. Your safety is the most important factor. Equipment can be replaced, so let go of any nets or other equipment and focus on getting your head above water and making it to shore safely.
8. You will be working outside during the lab. Be prepared for any type of weather (hot or cold, sunny or cloudy). There are no restrooms, so plan accordingly.

Summary Statistics and Excel

Several labs will require you to calculate summary statistics and growth rates, and to make graphs displaying those results. Although all of these processes are straightforward to do by hand, it becomes tedious when you are performing the same calculation on multiple data sets. There is also a higher likelihood for mistakes, and it is much harder to track down the source of the mistakes. Spreadsheet software like Excel makes these types of tasks much easier. If this is your first time using Excel it might be challenging, but with practice you can become quite skilled. Your TA will help you with the details of using Excel. This appendix details how to use Excel as a tool. If you get stuck on a particular step, ask for help from your TA, see the resources listed at the end of this lab, or search for an answer online.

Summary Statistics

Some of the most basic questions in science involve asking if a particular factor causes a response. For example, does adding more fertilizer increase plant growth? Does decreasing poverty decrease birth rates? Research involves designing experiments, testing hypothesis, and analyzing data to answer such questions. During this process, the researcher needs to determine if the results are due to the treatment (e.g., the extra fertilizer), or are they are due to chance and/or random variation. There is always some degree of randomness in experiments. Using the plant and fertilizer experiment as an example: Assume that we had two plants, and we gave one water with fertilizer (our experimental treatment) and the other just water (our control treatment). The plant with fertilizer grew 6 cm taller than the control plant. It is tempting to conclude that fertilizer increases growth, but what else could have caused these results? Maybe the plant that was given fertilizer was also older, and would have grown more even without the fertilizer. Maybe the control plant was diseased, or simply had a genetic makeup that produced shorter plants. Although a well-designed experiment tries to control as many of these potentially confounding effects as possible, if a researcher only tests one plant in each treatment, it will be very difficult to determine if the results were due simply to random factors.

The common solution is to use multiple replicates for each treatment, ideally assigned to each treatment at random. If all of the replicates in a particular treatment generally respond in the same fashion, then that increases our confidence that the experimental factor is what is causing the response. If there is no consistency in the results across the replicates, then perhaps that treatment does not affect the response. Overall, including multiple replicates (at least three) of each treatment increases our confidence in the conclusions we can draw from the results.

One challenge in using replicated treatments is that we are generating more data, which in turn makes it more difficult to understand and interpret the results. To help we can calculate summary statistics of the results. These generally include the location of the data (where most of the data are found), such as the mean (average), the median (midway point between lowest and highest value), and the mode (the value that occurs most often), and a measure of the spread of the data (how variable the data are), such as the standard deviation (how much each measurement varies from the mean) and the range (the span between the lowest and highest values). It is rare to see a measure of location without some measurement of the spread. We will use the *mean* as our measure of location and the *standard deviation* (SD) as our measurement of spread around the mean. The spread is often called the margin of error, which implies that the researcher made some mistake. This is possible, but not always the case. As mentioned earlier, there is natural variability in most systems that can influence the value of the standard deviation. Replicates that have similar values lead to low standard deviations, while those with a wide range of values have higher SDs (Fig. A2. 1). If the SDs among two or more treatments overlap, we generally conclude that those treatments are not significantly different from each other, even if the averages are different (a quick rule of thumb for estimating statistical significance). This basic problem of determining if treatments differ from each other due to the treatment effect and not just because of random variation is why researchers use statistical tests. We are not going to conduct a statistical analysis in our labs, but be aware that is the method for making formal conclusions about results.

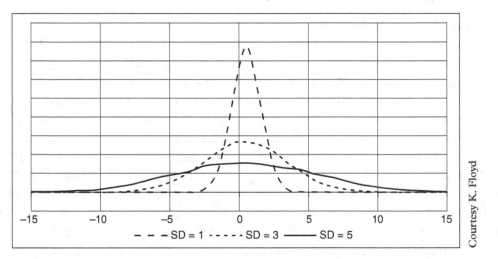

Figure A2. 1. The distribution of data with low, medium, and high variation as measured by standard deviation (SD). All three lines have a mean value of 0. The blue line shows a SD of 1, the orange line a SD of 3, and the gray line a SD of 5. When the SD is 1, all of the values are from approximately −3 to 3. When the SD is 5, there is a much greater range of values, from approximately −10 to 10. If measurements of several replicates in a particular treatment are very similar, the SD should be small.

Using Excel

Calculating the mean of a set of numbers is straightforward: add up the values and divide by the number of counts (replicates). Calculating the SD is a bit more complex, but not that difficult to calculate by hand. However, doing either calculation for larger data sets becomes time consuming very quickly. This is where spreadsheet software like Microsoft Excel or Google Sheets perform well. We will use Excel for recording our data and performing calculations (e.g., averages, SDs, and the growth rate calculations in the algae biofuel lab). Many of the difficulties in using such programs revolve around getting the data in the correct format to make the calculations. We provide templates on Blackboard to give examples of how to format data for each lab. The templates are not the only way to set up the data, but they are reasonably efficient. If you are going into a research or business field, you will probably be using spreadsheets. Even if you do not think that you will use spreadsheets in your career, they are a convenient method to track spending, make schedules, or anything that involves keeping information organized. The more you use spreadsheets, the more you will come to appreciate them.

We will also use Excel to make the graphs of data in some labs. Many of the difficulties of drawing graphs by hand, such as trying to determine the correct scale of the axes and figuring out where particular data points should be plotted are easily handled in Excel. We will be making two types of graphs during the semester: scatter plots and bar plots (called "column" plots in Excel). Scatter plots are appropriate when looking at how two variables in a set of data are related (Fig. A2. 2). Examples include how plant height and fertilizer amounts are related (Fig. A2. 2) or how population growth and time are related. Bar plots are commonly used when the treatments are categories, such as vehicle types (Fig. A2. 3). There is not an inherent way to order vehicle types: cars, trucks, and SUVs might be equally valid as trucks, cars, and SUVs. When plotting data that have replicates in each treatment, we will plot the mean with error bars that represent the SD values (Figs. A2. 2B, 3B).

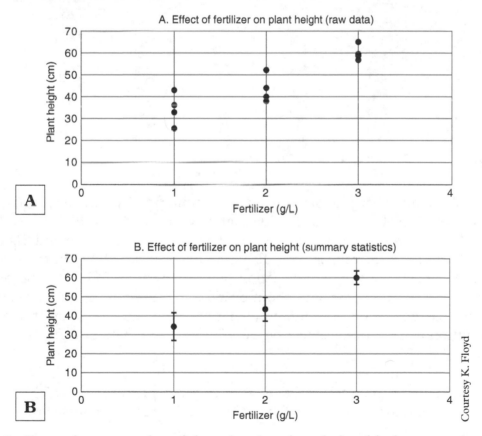

Figure A2. 2. Example scatter plot of data showing the relationship between the amount of fertilizer and the height of a plant. (A) A scatterplot of the raw data, showing values from all replicates per treatment. (B) A scatterplot of same data as in A, but using the summary statistics, showing the mean and SD (represented by the error bars) of the replicates for each treatment. Examine how the variability in the replicates shown in A impacts the size of the error bars in B. Using the rule of thumb that treatments with overlapping SDs are not likely to be different (if tested using formal statistical tests), which amount(s) of fertilizer lead to different plant heights?

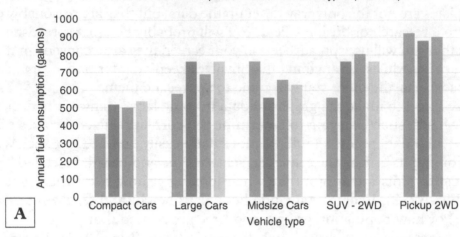

A. Annual fuel consumption of different vehicle types (raw data)

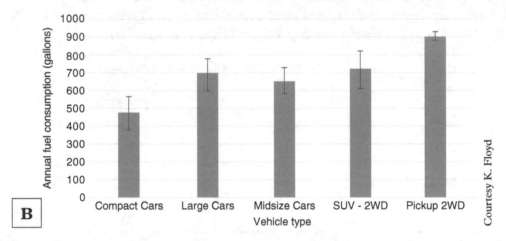

B. Annual fuel consumption of different vehicle types (summary statistics)

Courtesy K. Floyd

Figure A2. 3. Example bar plot of annual fuel consumption by vehicle type. (A) The annual fuel consumption of each vehicle is plotted separately, grouped by vehicle type. (B) The same data used in A but with summary statistics calculated to show the mean and SD (represented by the error bars) for each vehicle type. Examine how the variability in the individual data presented in A impacts the standard deviation values shown in B. What are the general trends in the data? Using the rule of thumb about overlapping standard deviations suggesting that the treatments/categories are not different, what can you conclude about the differences in fuel consumption across these vehicle types? Why are these data not presented as a scatter plot?

The instructions for how to use Excel vary a little depending on which version you are using. The main functions that you will be using Excel for in this course are:

1. Using formulas (mean, standard deviation, growth rates)
2. Sorting data
3. Making bar plots
4. Making scatter plots
5. Adding error bars to plots

You can readily find instructions on each of these through simple internet searches. There are some links to useful websites in the resources section. The Technology Support Center at UTEP also offers workshops on Excel. Their information is also given in the resources section.

Review the online instructions for each activity given in the resources section.

Using formulas

1. Click on the cell that you want the calculation or formula result
2. Enter "=" before the calculation or formula
3. The function for the mean is "average"
4. The function for the SD is "stdev"
5. The function for the natural log (used in algal growth calculations) is "ln"
6. For selecting a range of cells (like the replicates for a particular treatment) that the function will be applied to, enter the function, then an open parenthesis "(", then click and drag on the cells, then either type a closed parenthesis ")" or just hit enter. For example, to calculate the mean, you would type "=average([click and drag to select data cells])"
7. You can copy and paste formulas to save time, but be careful that the cells are arranged in the same way for each column or row that the formula is calculating

Sorting data

Excel will only sort the data that are selected. If you only select one column of a data set, and sort the data, only the column selected will be sorted. This will likely undo the relationships among the data. Be sure to select the entire data set before performing the sort function. Remember that you can always undo the command to restore the data to its prior order.

Plots

Once the data are in the correct format, select all the data and go to the "insert" tab and select the type of chart that you want. You can play with some of the options, but generally you should use straight lines on the scatter plots. Scatter plots allow both axes to scale according to the actual data, while line graphs treat the x-axis data as categorical and are evenly spaced.

Excel makes initial assumptions about which data should be in the rows and which should be in the columns, but often does not get it correct. One of the first troubling shooting steps to be taken if your graph does not show the relationships you intended is to click on the "switch row/column" button in the Chart tools -> design tab.

If there are missing data (perhaps your group forgot to measure one of the replicates), enter "#N/A" in that cell. This allows the remaining points to be connected in graphs.

Error bars

You first need to calculate a measure of spread (SD) for each treatment. Error bars will be under "more options" then "custom." This allows you to specify the exact values for the error bars. We will be using the same value for plus and minus (also called positive and negative in some menus), so our error bars will represent the value of the SD greater than the mean and the value of the SD less than the mean. Be sure that you are adding the SDs that correspond to for the particular treatment. When you add the error bar, Excel will ask which series it should be based on. Chose the appropriate treatment, and then once the error bars are included for that treatment, repeat for the remaining treatments.

Excel often includes horizontal error bars by default. Because we do not have any variability in the x-axis for our types of data, you should delete those error bars by selecting them by clicking on them and then hit the delete key. Repeat for all of the treatments.

Resources

The Technology Support Center at UTEP offers workshops on Excel (along with other programs). You might want to attend a session if you have never used Excel before. The schedule is found at http://admin.utep.edu/Default.aspx?tabid=74112.

Online Excel instructions (mostly using Excel 2013)
Instructions on basic use of formulas and calculations:
> http://fiveminutelessons.com/learn-microsoft-excel/how-enter-basic-formulas-and-calculations-excel
> http://www.excel-easy.com/introduction/formulas-functions.html

Short video from Goodwill Community Foundation (GCF): https://www.youtube.com/watch?v=BY8nX0CLIpU

Instructions on calculating means (averages):
http://spreadsheets.about.com/od/excelstatisticalfunctions/ss/Finding-the-Average-Value-with-Excels-AVERAGE-Function.htm

Short video from O'Reilly, also shows how to copy and paste formulas: https://www.youtube.com/watch?v=yCvi0RxkzmY

Instructions on calculating SDs:
> https://support.office.com/en-us/article/STDEV-function-51fecaaa-231e-4bbb-923`33650a72c9b0

Short video from StatisticsHowTo: https://www.youtube.com/watch?v=wJGgZJNYaPA

Instructions on sorting data:
http://www.contextures.com/xlSort01.html

Short video from GCF: https://www.youtube.com/watch?v=KS9N4yAjuYQ

Instructions on making column plots (in Excel, column plots are when the bars are perpendicular to the x-axis, and bar plots are when the bars are parallel to the x-axis):
> http://www.brighthub.com/computing/windows-platform/articles/128667.aspx

Short video from StatisticsHowTo: https://www.youtube.com/watch?v=dwaoJYSiTZ0

Instructions on making scatter plots:
> http://www.dummies.com/how-to/content/how-to-create-a-scatter-plot-in-excel.html

Short video from StatisticsHowTo: https://www.youtube.com/watch?v=0VtUQLbfewU

Little bit longer video showing multiple lines on the same plot, similar to what you will do in the algae biofuel labs: https://www.youtube.com/watch?v=_dB-EwQ9EhM

Instructions on adding error bars

Make sure that you follow instructions for *custom* error bars and add them based on the SD calculations already performed.
> https://nathanbrixius.wordpress.com/2013/02/11/adding-error-bars-to-charts-in-excel-2013/

Short video by Jeff Muday: https://www.youtube.com/watch?v=AfAG61UWsWA

Water Chemistry and Glassware Cleaning

General Instructions

Labeling glassware

Glassware includes beakers, flasks, graduated cylinders, centrifuge tubes, and glass vials, along with any other container for holding samples or solutions. Items called "glassware" can be glass or plastic. We will use Sharpies to write directly on most of the glassware. The ink can be removed using ethanol (EtOH).

Avoiding contamination

When collecting water samples from the environment (e.g., the Rio Grande), first rinse the collection container three times with sample water. We use buckets for the river samples, so fill the bucket up and dump it out several times before collecting the sample. If you are collecting a sample while standing in the river, collect the sample from upstream of you to avoid any contamination from your waders or sediment stirred up while walking (which will flow downstream with the current). If collecting samples from the algae bioreactor, first mix the solution to ensure a relatively homogeneous distribution of algae before collecting the sample.

After using a disposable pipette (plastic transfer pipette or serological pipette), throw it away or place in the recycling container instead of trying to rinse it for reuse. Although it is more wasteful to use disposable items, these types of pipettes are difficult to clean and thus present a significant risk of contaminating a future sample.

Waste disposal

The chemicals used in some water chemistry measurements (nitrate, phosphate, potassium, dissolved oxygen [DO], and alkalinity) must **not be poured down the sink drain**. There will be plastic waste containers available for all labs that include water chemistry measurements. Containers will be labeled for the specific types of waste they contain. Pour samples that contain chemical reagents into the correct waste container, and rinse the glassware that held the sample three times with distilled (DI) water, pouring the rinse water into the waste container. Clean the glassware as described next.

Cleaning glassware

1. Remove any labels using 95% EtOH. If you used 250 mL centrifuge bottles, do not remove the bottle and cap numbers or the mass of the bottle. If the numbers or weights are becoming hard to read, please rewrite them using a Sharpie.

2. Most glassware needs to be cleaned using the special glassware soap (in labeled squeeze bottles near the sink) and a brush, rinsed four times with tap water (the shiny tap on the left), then rinsed four times with distilled water (the dull tap on the right). Leave any caps or lids off, and either leave the glassware in the drying tubs on the center lab bench or in the smaller tubs that remain on the individual lab benches (your TA will tell you which location).

3. The exceptions to using soap are the materials dedicated to nutrient measurements (nitrate, phosphate, and potassium). The small glass test tubes, plastic nitratest tubes, plastic crushers, and the test tube rack (Fig. A3.1) only need to be rinsed four times with distilled water. The plastic nitratest tubes do not need to be perfectly clean. Just rinse as much of the sticky powder out into the appropriate waste disposal container as possible using the squeeze bottles of distilled water. Let your TA know if there are residues or scratches on the glass test tubes. Leave the caps off the tubes, and leave the glass tubes in the rack so they do not get broken by other supplies.

Cleaning the glassware is extremely important! Leaving soap residues or the original sample in the glassware will impact the results of the next person to use the glassware.

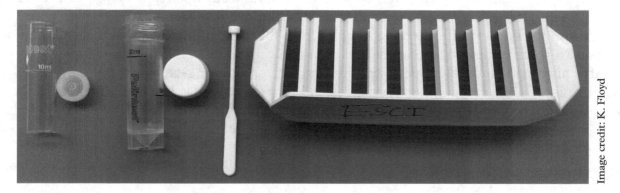

Image credit: K. Floyd

Figure A3.1. The glassware used in the nutrient measurements. From left, glass test tube, plastic nitratest tube, plastic crusher, and test tube rack. Only write above the 10 mL line on the glass test tube.

Measuring Nutrients

We will measure the concentrations of nitrate and phosphate several times (Algae Biofuels, River Field Trip, and Soils). We will also measure potassium during the Soil lab. These are three important nutrients for plant growth. We will also measure the amount of DO and the alkalinity of water samples during the River Field Trip. Nitrate, phosphate, and potassium measurements are colorimetric tests. The reagents added to the sample react to create a color, and the intensity of the color is proportional to the amount of the nutrient present. We measure the intensity of the color using a photometer. The DO and alkalinity tests also use chemical reactions that change colors, but we use the titration method (the slow and measured addition of a reagent) to determine the amount of DO or alkalinity. Detailed instructions about how to perform each measurement are available in the laboratory and posted on Blackboard. Some general instructions follow.

Colorimetric Tests

The photometer can only measure the color intensity within a given range. We will need to dilute the sample if the concentration of the nutrient is higher than what the photometer can measure. There are four basic dilutions that we commonly use in the lab (Table A3.1). You will always dilute the samples with distilled water, and it is critical that you write down which dilution factor you used. You will be measuring samples with unknown concentrations, so you might need to repeat a measurement if you initially did not dilute the sample sufficiently, or diluted it too much. This is not a mistake, but simply how research proceeds. Multiply the reading from the photometer by the dilution factor to determine the concentration in the original sample.

Table A3.1. Dilution factors for colorimetric tests. The amounts listed assume a final volume of 20 mL. If the test requires a final volume of 10 mL, use half of the listed amounts for both the sample and the DI water.

Dilution factor	Amount of sample (mL)	Amount of DI water (mL)
1	20	0
2	10	10
10	2	18
20	1	20

Only label the glass test tubes for the photometer at the top, above the 10 mL line. Clean the outside of the test tube (below the line) with 70% EtOH and dry with a clean tissue before placing the test tube in the photometer. Double check that the settings are correct for the nutrient you are measuring.

Titration tests

The most common mistake when doing a titration-based method is to add all of the titration reagent at once, instead of adding a few drops at a time. You will need to repeat the measurement if you added too much reagent. The titrator is only to be used for the particular titration reagent, not for transferring sample or other chemicals.

Overall, pay attention to what you are doing and follow the instructions for the specific measurements that you are performing. Clean glassware well so that it is ready for the next person to use. The people using the supplies before you did the same for you.

Microscope Instructions

Objectives

1. Identify the parts of a microscope and know their function
2. Demonstrate proficiency in the use of a microscope by being able to focus on objects under both low and high power using appropriate lighting
3. Determine total magnification
4. Distinguish between dissecting and light compound microscopes, and describe the strengths and weakness of each for various purposes

Introduction

Microscopes help us see objects that are too small to be seen with the naked eye. This is accomplished by increasing both *magnification* and *resolving power*. Magnification makes an object appear larger than its normal size, and resolving power (or resolution) is a measure of the clarity of the magnified object. There is almost no limit to how much magnification can be accomplished with a microscope, but the clarity of the image is limited by the wavelength of visible light used to illuminate the object. One aspect of clarity is the ability to distinguish between two objects when they are close together. With the naked eye, two objects must be 0.1 mm apart to appear as two objects, while they must be 0.2 micrometers apart using a light microscope (about 1000× closer than with the naked eye). If they are closer, they appear to be a single object.

There are a number of different types of microscopes that allow magnification at different levels.

Dissecting Scope	Very low magnification good for looking at small aquatic animals (zooplankton) or soil texture. Similar to a magnifying glass in terms of degree of magnification.
Light Compound Microscope	Moderate magnification. Used for identifying gross characteristics of cells, counting cells, observing their shape. Nucleus, nucleolus can be seen.
Electron Microscope	High magnification. Allows visualization of organelles within the cell.

The microscopes used in this lab are light microscopes, specifically **dissecting** and **compound microscopes** (Fig. A4.1). There are two lenses used in combination to increase magnification. The **ocular lens** is closest to the eye. The ocular lens usually enlarges the image 10 times (10×). The **objective lens** increases the magnification even more. There are often three objective lens, each with a different magnification (×4, ×40, and ×100). Total magnification or total enlargement is equal to the magnification of the ocular lens <u>multiplied by</u> the magnification of the objective lens.

Example:	Ocular	Objective	Total magnification
	×10	×4	×40
	×10	×40	_____
	×10	×100	_____

Care, transport, and set up of microscopes

1. When carrying a microscope, always put one hand under the base and the other hand on the arm of the microscope. Carry it upright, because there are parts that can fall off the microscope if it is tilted too far.
2. Look at the microscope and identify the parts of the microscope using Table A4.1 and Figure A4.1 as guides.
3. Plug the microscope into the outlet and adjust the light intensity to midrange. Adjust the distance between the ocular lenses to fit your eyes. You should be able to use both eyes when looking through the ocular lenses.

Table A4.1. Parts of the compound microscope (see Fig. A4.1).

Part	Function
Ocular lens or eye piece (1)	Provides ×10 magnification
Nose piece (2)	Holds objective lens
Objective lens (3)	Three lenses: provides ×4 (low), ×40 magnification and ×100 with oil immersion (Note that other microscopes might have objectives with different magnification, but the magnification is usually labeled on the side of the lens)
Course focus knob (4)	Used to do initial focusing of slide
Fine focus knob (5)	Sharpens image after initial focusing
Stage with clips (6)	Holds slide to stage
Light source (7)	A built in electrical light (illuminator)
Iris diaphragm (8)	Regulates the amount of light passing through the slide
Mechanical stage control (9)	Supports and moves slide
Condenser	Focuses light from illuminator into a beam that passes through the slide

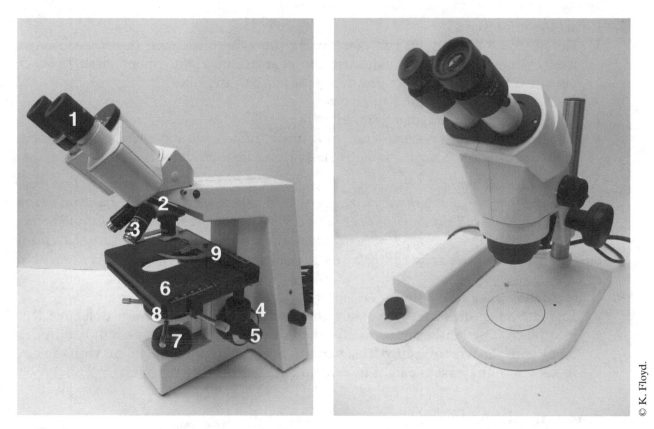

© K. Floyd.

Figure A4.1. Light compound microscope (left) and dissecting microscope (right). Refer to Table A4.1 for the names and functions of the labeled parts. Several of the parts are the same in both microscopes, so you can use the labeled figure to identify the parts of the dissecting microscope.

Operation of dissecting microscopes

1. Pour a small amount of a water sample into a petri dish.
2. Place petri dish (or specimen of interest) on the microscope stage.
3. Adjust the magnification to the lowest power by turning the knob counter clockwise.
4. Focus on the edge of the dish, and then move the specimen to the center of the field of view.
5. Increase magnification to see details of the organisms.
6. Return to the lowest level of magnification.

Operation of compound microscopes

1. Place slide on the stage (see following instruction for preparing a wet mount slide).
   - Rotate objective lens out of the way.
   - Secure slide with stage clips.
   - Move specimen into light path using mechanical stage controls.
   - Rotate objective lens back into position, starting with the lowest power lens.

2. Focus the microscope.
   - Rotate the ×4 objective into place using the nose piece ring (not the objective itself). Looking from the side, use the coarse focus adjustment knob to place objective lens at lowermost position (lens close to the slide).
   - Move condenser to topmost position.
   - Close the iris diaphragm about half-way.
   - Looking through the ocular lens, bring object into rough focus using coarse focus adjustment knob. Sharpen image using fine adjustment knob.

   (*Hints: If you can't find the image, start over. Go back to the lowermost position and move the focus up very slowly with the coarse focus knob. Small objects should come into focus within one or two full rotations of the coarse adjustment knob. It may also help to reduce the amount of light, especially for clear and/or colorless specimens. In some cases, focusing on the edge of the cover slip will help you get close to the correct depth of focus for your object.*)

3. Increase magnification.
   - Increase magnification by rotating the next higher-magnification objective lens into place. The lenses are *parfocal*, which means that the object should still be in focus with the next objective lens, perhaps needing minor adjustment with the fine focus knob. Increase the light by opening the iris diaphragm.

Shut down of microscopes

1. Remove specimen from stage. Clean stage with Kimwipes or paper towels. Clean lenses with **lens paper** (do not use any other type of material on lenses because they can scratch).
2. Rotate lowest power objective into place using nosepiece. Put objective lens into lowermost position with coarse adjustment knob.
3. Turn off light. Unplug microscope and wrap cord around arm.
4. Cover. Place on shelf, if instructed to do so.

Preparation of a wet mount

A wet mount is a way to view cells suspended in liquid (most often water or saline). It can be used to view most types of cells, including (but not limited to) bacteria, fungi, animal tissue, plant tissue, and protozoans.

1. Place a drop of liquid (water, saline, pond water) on a glass slide.
2. Place the specimen into the liquid. This could be bacteria, a piece of a leaf, algae culture, tissue cells, and so on. Mix, if needed, with toothpick, dissecting needle, edge of coverslip, and so on. Mixing may be needed to break up a large cluster of organisms or to release organisms from sticking to the surface film of the water drop.
3. Lower a coverslip over the drop. Touch one side of coverslip to edge of drop (about a 45° angle), then lower slowly. Try not to trap air bubbles. If viewing pond water, scrap a small amount of clay onto the corners of the cover slip before lowering it over the

drop. This creates a little extra space that prevents any larger organisms from getting compressed, distorted, or killed.

4. View the specimen. The challenge of viewing a wet mount is getting the light right. Many cells (and small organisms) are colorless and fairly transparent, and when suspended in a clear medium, they are difficult to see. For best results, start off with illumination very low. Look for the outline of the cell and then increase light as needed. If you are having trouble locating cells and organisms, start at the edges of the coverslip where cells and organisms often accumulate.